THE DAWN OF INDUSTRIAL AGRICULTURE IN IOWA

THE DAWN OF INDUSTRIAL AGRICULTURE IN IOWA

Anthropology, Literature, and History

E. Paul Durrenberger

UNIVERSITY PRESS OF COLORADO
Louisville

Published by University Press of Colorado
245 Century Circle, Suite 202
Louisville, Colorado 80027

 The University Press of Colorado is a proud member of the Association of University Presses.

The University Press of Colorado is a cooperative publishing enterprise supported, in part, by Adams State University, Colorado State University, Fort Lewis College, Metropolitan State University of Denver, Regis University, University of Alaska Fairbanks, University of Colorado, University of Denver, University of Northern Colorado, University of Wyoming, Utah State University, and Western Colorado University.

∞ This paper meets the requirements of the ANSI/NISO Z39.48-1992 (Permanence of Paper)

ISBN: 978-1-64642-206-7 (cloth)
ISBN: 978-1-64642-207-4 (paperback)
ISBN: 978-1-64642-208-1 (ebook)
https://doi.org/10.5876/9781646422081

Library of Congress Cataloging-in-Publication Data

Names: Durrenberger, E. Paul, 1943– author.
Title: The dawn of industrial agriculture in Iowa : anthropology, literature, and history / E. Paul Durrenberger.
Description: Louisville : University Press of Colorado, [2021] | Includes bibliographical references and index.
Identifiers: LCCN 2021029517 (print) | LCCN 2021029518 (ebook) | ISBN 9781646422067 (cloth) | ISBN 9781646422074 (paperback) | ISBN 9781646422081 (epub)
Subjects: LCSH: Corey, Paul, 1903–1992—Correspondence. | Lechlitner, Ruth, 1901–1989—Correspondence. | Agriculture—Iowa—History. | Agricultural industries—Iowa—History. | Farm management—Iowa—History. | Land use, Rural—Iowa—History.
Classification: LCC S451.I8 D87 2021 (print) | LCC S451.I8 (ebook) | DDC 338.109777—dc23
LC record available at https://lccn.loc.gov/2021029517
LC ebook record available at https://lccn.loc.gov/2021029518

Some material in this book was previously published and is used with permission: E. Paul Durrenberger, "Ethnography, Naturalism, and Paul Corey," Anthropology and Humanism 44, no. 1 (2019): 127–143.

Cover illustration: Grant Wood (American, 1891–1942), Stone City, Iowa, 1930, oil on panel, Joslyn Art Museum, Omaha, Nebraska, Gift of the Art Institute of Omaha, 1930.25. 2021 Figge Art Museum, successors to the Estate of Nan Wood Graham/Licensed by VAGA at Artists Rights Society (ARS), NY.

To

Suzan Erem

and

The Sustainable Iowa Land Trust:
Dedicated to protecting Iowa land to grow
healthy, clean table food regeneratively

and

*To all those who love the American people and believe
in their capacity to create plenty for all.*
Anna Louise Strong, dedication in *My Native Land* (1940), 3

We had no second sight, and none of us could imagine the years that were to come. It was a moment when our country was grand and splendid beyond what I can make a reader imagine today, regardless of how eloquent I may be.

Howard Fast, Being Red *(1990), 14*

Who shall own and command America—the people or the profiteers? This is the great question of our time. This is the question whose answer will affect the course of world history.

Anna Louise Strong, My Native Land *(1940), 299*

CONTENTS

ACKNOWLEDGMENTS

First and foremost I thank my wife, Suzan Erem, for her support during the four years I've been working on this book; for listening to me rant about the Depression, the thirties, the twenties, the New Deal, Paul Corey, Ruth Lechlitner, the Farm Bureau, and all of it, for listening to my musings and asking me pointed questions.

My colleague Dimitra Doukas and I have carried on a correspondence, an epistolary relationship worthy of Paul Corey and his correspondents. She helped me frame questions and offered penetrating insights. I thank my colleague Barbara Dilly, a native Iowan who knows rural Iowa from the inside out. She has taught me much about Iowa and offered me valuable comments on early drafts of the book.

I owe a great debt to David Plath, the professor at the University of Illinois who first started me thinking about questions of fiction and anthropology when I was a graduate student there.

I thank Charlotte Steinhardt of the University Press of Colorado for her patience and her faith in this unorthodox project.

Many people whose names I do not know provided essential assistance. I thank the staff of Special Collections at the University of Iowa Library where I spent untold hours reading and pondering. But especially I thank them for the use of their scanner, which allowed me to convert important materials to digital formats so I could reference them at home while I was writing. I had many occasions to appreciate the efforts of the unnamed archivists who organized the papers. The anonymous readers that the University Press of Colorado recruited provided useful comments and suggestions on earlier drafts of this book.

THE DAWN OF INDUSTRIAL AGRICULTURE IN IOWA

INTRODUCTION

Questions

The history of the United States is the history of capitalism, the transfer of wealth from those who create it to those who control it. The history of its culture is a history of ways to either disguise that process or to normalize it and make it acceptable. This dynamic was established early in the Northeast and spread west as America's robber barons incorporated the hinterlands as suppliers of raw materials.[1] In the mid-nineteenth century, Iowa was at the western edge of that movement. The railroads and their industrial infrastructure moved Iowa to the center in a transformation from the original family farms of the settlement period into today's agricultural industry. This book is about these developments as a part of America and about two Midwesterners who chronicled them as Americans.

Paul Corey was born in western Iowa at the beginning of the twentieth century; Ruth Lechlitner was born in Indiana about the same time. Paul graduated from the University of Iowa in 1925, and Ruth received an MA in 1926. Their lives spanned the most consequential period in history of the

DOI: 10.5876/9781646422081.c000

United States, one that is difficult to imagine from the perspective of a hundred years later.

Paul and Ruth told the stories of these processes in poetry and fiction. How did these remote political and economic forces shape their lives? How did Paul and Ruth depict those forces? And finally, what are the relationships of fiction, individual lives, and anthropology?

The lives of Paul Corey and Ruth Lechlitner offer us a window into life between 1925 and 1947, one of the most radical and creative in the history of the United States, a period when everything was in the balance, when it appeared the country could as easily move toward fascism as toward communism, a period before the warfare state and the cultural and political repression that was to engulf them.

Iowa's Industry

Today Iowa, the heartland of the United States, is covered by a horizontal industry that is as polluting as the old vertical ones of the Rust Belt. The oft-repeated slogan, "we feed the world," is true to the extent that this industry produces the ingredients of the American diet that are responsible for our epidemic of obesity and diabetes and associated health problems. This industry also controls the political structure of the state.

If industrial farming is so bad, so damaging, why is it still going? Why haven't the perpetrators stopped? Having grown up on an industrial farm, and having a father who still farms that way, agricultural journalist and writer Stephanie Anderson is very accepting of industrial farmers. Many farmers could make the switch from damaging industrial to helpful restorative farming but do not because they are trapped in the industrial system, running from one season to the next on the productivity treadmill—borrow money, buy inputs to improve productivity and acquire more land, harvest more, watch prices fall, repay as much of the loan as they can. Or, if prices go up, acquire more land and bigger equipment. And the accompanying debt. And repeat. Year after year.[2]

So good people are trapped in a bad system that is contributing to the destruction of our planet and our health. But, Anderson says, there comes a time "when claiming to be a good person trapped in a bad system is no longer an excuse for inaction—and that time is right now."[3] We know organic regenerative agriculture can feed the world and support farm families. We also know that industrial agriculture doesn't work for our society or our environment. What's the holdup?

Those masters of the ruggedly independent American farmer, agribusinesses and their allies, are very powerful and control national and state governments and their farm policies. That's the reason. One means by which they do this is by enacting subsidy programs that keep the prices of American food at artificially low levels that do not account for the costs of environmental damage or harm to our health, either through eating their products or sharing the planet with them. In this they are powerfully aided and abetted by the universities created by the Morrill Land-Grant Acts and the extension programs that provide the corporate research and ideologies to justify and promote industrial farming.[4]

I use the term "industrial" farming rather than "conventional" because, as Anderson suggests, "conventional" sounds like it's the "way things are supposed to be."[5] That's no mistake. The industrial agriculture movement is famous for its Orwellian twists of language to make "water pollution" "water quality," and chemical pollutants and swine sewage "nutrients," and "we feed the world" for "we destroy our planet," just to name three examples.

So, yes, those ruggedly individualistic farmers are not feeding the world—they're destroying it to serve their agribusiness masters. But Anderson emphatically argues that the "good people in a bad system" excuse will no longer hold up. It's time to grow a spine and get out of that system and follow the lead of people who, like the farmers she visited and wrote about, are making their livings by practicing regenerative agriculture to restore rather than destroy the planet.

Journalist Eric Schlosser starts his story with fast foods and shows how they are related to virtually every dimension of American life from transportation to politics and housing.[6] Our story starts from the other end, the production of the components of those fast foods. It gets to the same dire place of pollution, political corruption, and epidemiology all centered on the heartland of Iowa, a place cloaked in a culture of niceness that prohibits the mention of such unpleasantness.

This is a complex of practices and ideas as integrated as the culture of the Plains Indians that preceded it and just as amenable to a cultural ecological understanding.[7] Both understandings rest on often contradictory ethnography.[8] A system this complex offers many structural positions, each with its own view of the world; none may encompass the whole system. Some in the system actively attempt to conceal its operation. Ethnography can describe this system from the various points of view, but to understand the system as a whole, we have to step outside of it and grasp the totality. Paul Corey's classmate in the journalism department at the University of Iowa, George Gallup (1901–1984), grappled with this problem and developed the method of polling as a means of characterizing whole social orders statistically. Sociologists often use government statistics to develop such descriptions and supplement them with their own surveys. Anthropologists rely on ethnography, the experience of living with the people we want to understand and observing closely their daily lives. Each pixel contributes to the picture; none is the whole. This brings anthropologists face to face with the problem of deciding what to believe or how to encompass the various points of view with all of their contradictions. Privileging any single viewpoint provides a skewed vision of the whole. Paul Corey transcended the various perspectives in his fiction. That is where I begin.

Perspectives

Literary critics of the period provided contemporaries' insights into the fiction and poetry of the time. Literary historians have commented from a temporally alien perspective. American Studies scholars have commented on the commentaries. All told, there is an intimidating pyramid of literature cascading from the novels and memoirs of the period to the contemporary appreciation of them to the remote analysis of them to the examination and critique of those analyses. Layer upon layer, farther and farther from the original writers and events at each echelon until the realities of the period and the people are blurred under blankets of interpretations. What can an anthropologist contribute to that mélange?

Anthropologists like to stay close to the material we're trying to understand. So this is not a work of literary history or literary criticism. Nor is it ethnography. It is something else, an anthropological biography, the approach to a period via the lives to two people, the examination of the lives of two people to reveal something about the subject matter about which Paul Corey wrote: family farming in Iowa. Along the way it provides the context for their lives, the period in which Paul and Ruth were most active in writing, between 1925 and 1947, which encompassed the collapse of capitalism, the rise of the Soviet Union, the rise and defeat of fascism, the postwar Great Repression, and the creation of today's continuous warfare state in the United States.

Anthropology and Biography

From the point of view of anthropology, any single person is one node in a network extending back to our first ancestors in Africa, into the indefinite future and around the planet. There are no individuals; we are all instances of greater wholes. An individual can be a window on a whole culture or illustrate otherwise invisible relationships and structures. Illustrative of this was the last of the Native American Yana people, Ishi, who in 1911 walked out of the forest in California where the anthropologist Alfred Kroeber recorded his language and stories. His wife, Theodora Kroeber, wrote his biography, *Ishi in Two Worlds*. Similarly, in 1929 the anthropologist Oliver La Farge wrote *Laughing Boy*, a fictionalized biography of a Navajo to illustrate the encroachment of Euro-Americans. Or Ruth Behar's story of a Mexican woman.[9] More recently, there is anthropologist Gísli Pálsson's *The Man who Stole Himself* (2016) about a Caribbean slave of African and Danish descent who washed up in a small hamlet in nineteenth-century Iceland. The understanding of each of those lives contributed to the understanding of larger configurations and processes.

Life histories were part of the practice of early anthropology in the United States[10] when anthropologists followed the army into Native American lands to chronicle their lifeways.[11] Many of them, like Ishi's Yana, were decimated if not abolished. The remnants were driven onto reservations and reduced to dependency on the Bureau of Indian Affairs. By exploring the patterns of the lives of individuals, anthropologists hoped to reconstruct the living cultures of which they'd been parts. These exercises were based on the assumption that individuals do not stray far from the patterns of their upbringing.

Nor did the principles of this book, Paul Corey and Ruth Lechlitner. Nor can anyone. João de Pina-Cabral wrote that while anthropologists identify the material or mental conditions that frame events,[12] we must remember that human speech and action is underdetermined; that there is a lot of leeway in the actions of people, even if they are structurally determined. Author Howard Fast, for instance, wrote that while the experience of poverty and misery was burned into his soul, and nothing changed in the scheme of poverty, what did change was his ability to face and alter circumstances: "I ceased to be wholly a victim."[13]

Looking at the question from a different perspective, anthropologist Philip Carl Salzman summed up E. E. Evans-Pritchard's wartime study of a Libyan province by saying that it shows that "an understanding of culture and structure alone are not sufficient for explaining human destinies; as well we must take into account the agency and acts of the multiple parties and the events these generate."[14]

Paul Corey and His Fiction

Paul Corey was born on a farm in western Iowa in 1902, took a journalism degree from the University of Iowa in 1925, married poet Ruth Lechlitner, lived with her in France for some months, and went to homestead and write in rural New York. His ambition was to produce a multivolume series of interlinked novels as a social history to describe to the rest of the world rural life in Iowa and the death of family farming there. Three volumes of the series were published before the Second World War and a fourth just after. After the war, Corey was in the process of being disappeared during the Great Repression and knew it. He thus abandoned the remaining volumes of his project and moved to California where he and Ruth homesteaded again and where they died toward the end of the twentieth century. Here I sample the fourth volume of the series, *Acres of Antaeus* (1946):

> The roar of the crowd sounded like a gigantic corn-sheller thundering at its job, and Emily climbed to the top of the cab on the Winters truck to get a better view. Down Main Street all she could see was the blue overalls and jackets of farmers. Not a citizen of Markley was in sight. The farmers had taken over the town. The courthouse square was packed with them, jostling and shifting, talking, sometimes laughing and sometimes cursing. But the whole crowd was quick to sober.
>
> Sometimes angry shouts were flung up. "We want our farms! To hell with the insurance companies!" On one of the sandstone arms of the west

courthouse steps, Emily saw the unmistakable Stetson hat of Carleton Smith, the old Holidayer.[15] She heard his oratorical voice ranting and exhorting the crowd and the swelling roar of angry voices reached her like a blast of wind.

"What's happening inside?" she asked one of the farmers standing near the truck fender.

"Don't know, Miss. They been in there quite a spell now."

An hour had passed after the farmers crowded into the courtroom before Judge Hart came out of his chambers and sat down behind his bench. His face was gray but stern and stubborn. The bailiff stood at the edge of his desk with the foreclosure judgments in his hand, while the clerk sat down below chewing gum nervously.

As the farmers pushed in, the railing creaked and cracked under the strain. The blue-uniformed throng [of farmers] crowded toward the bench. Judge Hart rapped loudly for order. Someone back in the crowd yelled: "We want our farms back, Judge!"

"Order!" shouted the judge. "I'm, here to carry out my legal duties as Judge of Baldwin County . . ."

A voice interrupted him "Don't sign them judgements or you'll regret it."

"Are you threatening the court?" yelled the judge.

"We want our farms back."

"Order! Or I'll have to clear the court."

"We want our farms back." The demand became a chant. "We want our farms! We want our farms!"

The judge made a sign to the sheriff. "I order you to clear the courtroom."

Silence fell on the crowd. No one moved. The sheriff stepped forward and ostentatiously drew his gun. His lips twitched several times, but he said nothing. He took a step toward the crowded enclosure before the bench, and stood close to Al Winters.

Old Jeff stood opposite the officer, his lean face set, prepared to meet stubbornness with equal stubbornness. The judge could not sign the foreclosures in the face of all this pressure. He could not do it. The sheriff could not clear the courtroom. The farmers simply would not leave.

. . . The crowd surged back and forth, some trying to get out of the room, some rushing the judge and the sheriff. There were yells, hoarse shouts, and the sound of blows. The crash and splintering of the railing echoed through the other sounds.

Old Jeff saw the judge dragged over his bench, struggling and cursing. The sheriff lay unconscious on the floor. The bailiff, the clerk, and the lawyer for Corn-States fled through the chambers door. Old Jeff and Roy were swept along with the crowd out of the courtroom, the others dragging the grunting, fighting judge along with them.

Emily heard that revolver shot above the rising and falling roar of the crowd. An abrupt, terrifying silence fell upon the huge throng of farmers; then a murmur began. "What was that?" "Who's doin' the shootin'?" It grew pushing toward the courthouse. It seemed to squeeze on the square sandstone building from all sides. In a moment the answers came circulating back.[16]

This scene is a fictionalized version of events in Le Mars, Iowa, on April 27, 1933. The 1930s was a period of intense activism across Iowa, and Paul Corey's depictions of farm life in western Iowa provide a rich background to the primary and secondary sources on the period.

The 1930s in Iowa

Corey's Mantz trilogy—*Three Miles Square* (1939), *The Road Returns* (1940), and *County Seat* (1941)—were published by Indianapolis's Bobbs-Merrill after every New York press had rejected them. His *Acres of Antaeus* came out in 1946 from Henry Holt of New York. The publication of that volume also has a complex history. Those events of the radical thirties have been expunged from the consciousness of present-day Iowans and from the history of Iowa. Iowans can remember Herbert Hoover, a native of West Branch, Iowa, and celebrate him in an annual festival, though not by reenactments of Hoovervilles. But people do not remember or celebrate Paul Corey or any of Iowa's activists of the thirties, much less this scene from *Acres of Antaeus*.

Organization of the Book

Add to these overt political and less overt social exercises of power a great cultural revolution that had been underway since the turn of the twentieth century to remake American culture so that corporate rapacity would become acceptable and normal and we see a powerful engine for obliterating nonconforming events and people from memory.[17] It has been largely successful in shaping our sense of history so that we celebrate Hoover but not Hoovervilles and family farms where there are none. The first chapter discusses how Corey was expunged from Iowa's memory and how that was part of larger historical developments in the United States.

The second chapter tells the story that engaged Paul Corey's attention, the industrialization of Iowa's agriculture and the death of its family farms. The third is the story of the farm boy who went to the Mecca for writers, New York City, and how that experience convinced him of the necessity to free himself from the burden of a job in order to write. That was the chief reason he homesteaded—to make himself as independent as possible of the surrounding economy. This provided ample opportunity for him to live by his gospel of work, the old Puritan idea that work toward some objective is its own reward. This stands in opposition to the gospel of wealth that was being pronounced by Andrew Carnegie and his industrial cohort, who were intent to remake American culture into one that would appreciate and value their corporate rapacity.

Three broad movements are evident in the lives and writings of Corey and Lechlitner. The first was the movement to flee the ills of the industrial cities, a back-to-the-land movement of which they were a part and to which they contributed. I discuss this movement in chapters 3 and 4 as one branch of America's fascination with the rural. The other branch was the country life movement that promoted the industrial principles of efficiency and profits in an attempt to urbanize farmers. Corporate leaders such as Carnegie, who considered themselves to be the pinnacle of evolutionary processes that made them the natural managers of wealth for the good of the wider society, used their wealth to promote an ideology to justify their avarice and to found the discipline of economics that would add the seal of approval of science. The revolution to remake American culture in the image of this ideology is one theme of this book. Struggling against this tide were the back-to-the-landers such as Corey and Lechlitner.

The second was the literary movement of regionalism, associated in Europe with the rise of fascism and thus suspicious in the United States in this age of flux. Regionalism and its association with fascism are the topics of chapter 5. By virtue of his focus on Iowa farming, Corey was a regionalist. But since his fiction also encompassed California as well as rural and urban New York and Chicago, his subject matter was not confined to his natal region. There was, however, a strong regional movement that defined a distinct culture of the Midwest and heavily influenced Corey, not least in his dedication to the gospel of work.

This is the culture that Paul Corey was fleeing with no sense of nostalgia. The wide horizons of Iowa, Corey said, gave Midwesterners their frank, clear, and open eyes but beckoned to him, made him uneasy with an anxiety that led him beyond the horizon. Those thousands of miles of rolling plains were uninteresting and drab, "terrifying with stark, barren monotony."[18] Many never returned and never wanted to, he said. "I, for example, do not plan to ever return to live in Iowa. The vast gently rolling plains bore me, but the atmosphere of my youth, the perception of distant horizons, still haunts me."[19]

The perspective of distance allowed Corey to describe the details of daily life on Iowa farms in the first decades of the twentieth century in all of their sordid detail and to leave an ethnographic record of that time and place. He wanted to preserve a record of events and people that were being erased from consciousness and history.

The third movement was the stylistic movement of naturalism that Corey found best fitted to his subject matter, writing about the details of peoples' lives and the processes that determined them. It is a descriptive and gritty style associated in the United States with hard times and is still characteristic of American writing. In chapter 6 I discuss this movement and its relationship with ethnography, which exemplifies it. This opens the question of the relationship between fiction and ethnography, which has a long history in American anthropology.

Chapter 7 describes the overall plan that Corey developed for this project. In chapter 8 I discuss Corey's style of writing, his particular adaptation of naturalism to his subject matter. In chapter 9 I return to Corey's writing style as that was an important matter in his relationship with the publishing industry. He insisted against all advice on his collective style of writing that emphasized no heroes or villains, that had no plot or resolution, that entailed no judgment—that was in fact ethnographic and thus did not always fit the strictures of fiction. That sets the stage for his engagement with the publishing trade—the cultural apparatus. The correspondence in chapter 10 between Corey and various publishers regarding his story "Number Two Head-Saw" illustrates both his ideas of group-focused stories versus the more usual individually focused stories and editors' negative responses to the notion.

The correspondence of both the novelist and his poet wife Ruth Lechlitner shows the power of what sociologist C. Wright Mills called the cultural apparatus,[20] those networks of social relations that determine whose work gets displayed and whose is destined for oblivion. On the one hand, the editorial process guarantees a certain level of quality; on the other, it can squelch lines of inquiry, genres of writing, and politically fraught material. These days, that means forcing the creations onto the Internet; in the thirties, it meant oblivion. All of these were parts of larger cultural processes in the United States. This is the subject of chapter 9.

Chapter 10 provides an example of Corey's style of writing in a previously unpublished story. Paul was too old to serve in the Second World War, but he was intent to organize a militia that could meet what he thought was an imminent German aerial invasion of the eastern United States. This engaged him in a relationship with a wealthy and powerful industrialist. The correspondence between the two is a striking illustration of the differences between the gospel of work and the gospel of wealth, the topic of chapter 11. Chapter 12 treats the eve of the war, the looming sense of dread, and Ruth's treatment of it in poetry. It also deals with the evolving relationship between Paul and Ruth as she becomes pregnant a third time.

After the war the couple and their daughter move to California to repeat the homesteading experience. Chapter 13 treats the move, their later life, and their final days. Having come to the end of the stories of Paul and Ruth, the book devotes the final chapter, 14, to an epilogue. Iowa drops from their story when they move to California, and I do not treat their lives in California in detail.

Now, to anticipate the fate of Iowa's farms that I discuss in chapter 2, here I return to *Acres of Antaeus*.

Acres of Antaeus Again

In the end, the corporation wins and recedes into invisibility, engulfs all, even Jim, the protagonist. Jim starts the novel working as a foreman of a road-paving crew, meets school teacher Emily at a street dance in Markley, and takes a job at Mid-West Farms because "They're going to farm Iowa the way it'll be farmed in the future,"[21] with industrial equipment, management, and methods on land that insurance companies and banks have repossessed from failing family farmers. Emily, daughter of a farmer, disapproves of Jim's job with the heartless corporation, but he explains that he can change the corporation from the inside.

Jim visits with the Chief, one Mr. Granger, who explains mass production in agriculture and, with a gesture toward a map of the corporate holdings says, "We have our great industrial empires in America; well, I'm building an agricultural empire."[22] One of Jim's first tasks is to prepare for plowing by removing the fence between the Clausen and the Harris farms. Harris had just died of pneumonia. Clausen is still on his farm, which has been absorbed into the empire, and is none too pleased. Charley is a worker who hadn't seen two bits all winter and is glad of a chance to earn a buck doing whatever he's instructed, but as time goes on he begins to organize workers. At Christmas, Jim assesses his situation to Emily:

> "Last spring I thought I had a good job," he went on. "I thought Mid-West was doing things the way they should be done. I was farming in a big way. But my job isn't farming in a big way any more; my job's to get all the work I can out of poor devils and pay them as little as possible. I've got to hang on to that job because there isn't any other job to be had."[23]

Clausen dies of a heart attack; Jim visits, and the widow says blankly, "There's an end to everything."[24] Corporation thugs have beaten up Charley, the Red, who lies bandaged in the Clausen parlor as Otto Mantz vows to take care of the gang Mid-West sent to beat up Charley. Jim informs Mantz that the person who sent the thugs has been replaced as the corporation is now conscious of its public image. The corporation has hired the judge for $20,000 to induce him to drop all charges he might bring. Through such machinations, the corporation becomes invisible and erases any traces of resistance.

Jim concludes his visit to Charley and the widow at the Clausen's place and buttons his coat:

> With an uncertain gesture, he turned and walked quickly out through the kitchen. He drove out of the yard with a sense that the confusion in his life had begun to clear. Soon it would be time for spring sowing. The picture of tractors roaring down a field rose in his mind. He whistled a tune.
>
> As he sent his car droning along the open highway he turned things over in his mind. The days of building empires were over, even agricultural empires.

The land had to belong to the people who farmed it; the land was too much of a living thing to be left to the quarreling and bickering of men interested only in making money. The farmers would have to learn to own the machines of mass production co-operatively; then they could compete with the growing corporation farms.

Tonight, he decided, he would call Emily and tell her how the situation stood. . . . This year, he told himself, he'd see to it that the workers for Mid-West got a better break. When he swung in under the arch over the gateway, it was too dark to read the sign: "MID-WEST FARMS, INC."[25]

The corporation has faded into the dark, erased from view just as this period of history has receded from the awareness of Iowans and most Americans.

1

AMERICA'S DISAPPEARED

Nowhere have [writers] led such short lives as writers or yearned so
desperately to be in tune with the age or enraged against it.
   Alfred Kazin, On Native Grounds, *34*

To Disappear

The use of the word "disappear" as a transitive verb originated to describe
what happened to people who were out of favor with autocratic regimes. "She
was disappeared." She was among us one day and gone the next. No one could
learn where she might be. A prison cell? A concentration camp? A mass burial?
One official referred queries to the next to the next in a Kafka-like maze of no
exit. There was no formal, much less paper, trail to follow. All dead ends.

   When the regime changed and archaeologists and forensic scientists were
called in to excavate and identify the people in the mass burials, the long-
standing questions began to be at least partially answered.

   That hasn't happened in the United States. People have been lynched,
blacklisted, deprived of jobs and livelihoods, but nothing has happened on
the scale of the historical modern disappearances like those associated with
Stalin or Pinochet. But there are many American authors of the thirties, the
most radical period of recent American history, who have disappeared from
or never made it onto high school or college reading lists. Hemingway, Dos

DOI: 10.5876/9781646422081.c001

Passos, Virginia Wolf, Thomas Wolfe, Fitzgerald, Lewis, Stein. These we remember. But Dorothy Thompson, Agnes Smedley, Meridel Le Sueur, Anna Louise Strong, Josephine Herbst, Milly Bennett, Tilly Olsen, Nelson Algren, Alexander Saxton, John Reed, Vincent Sheean, Jack Conroy, Tess Slesinger, Malcolm Cowley, Granville Hicks?[1] Sometimes a biographer will rediscover a writer and bring her to our attention in a book, as Douglas Wixson did for Jack Conroy, Elinor Langer for Josephine Herbst, Tracy Strong and Helene Keyssar for Anna Louise Strong, or Tom Grunfeld for Milly Bennett.[2] The 1981 movie *Reds* resurrected John Reed. The feminist movement found Tillie Olsen in time to celebrate her before she died.[3]

Some breathed their last literary gasps in the sixties; some were at least briefly revived by the feminist movement, some by the rediscovery of radical writing.[4] Most were disappeared. Not in the Stalinist sense, but in a more insidious process of smothering, of being ignored out of existence. Most of the people I listed were communists or fellow travelers throughout most of the thirties when there was a viable context for a radical assessment of capitalism in America. But as that context waned, so did the possibility of critique, and with it, the authors who wrote from radical perspectives and their works.

Many left the Communist Party with the Russian-German nonaggression pact of 1939 or the Stalin show trials.[5] The first showed the USSR was no longer the bastion of anti-fascism; the second, that it could be as bad as the fascists. The last prewar act of the struggle between fascism and democracy was the Spanish Republic's from 1936 to 1939. Steadfastly "neutral," the United States outlawed those Americans who fought for the Republic in the legendary Lincoln Brigade that the communists organized.[6]

After the 1939 nonaggression pact between Germany and the Soviet Union, the US government increased its intrusion into the lives of citizens.[7] FBI director J. Edgar Hoover ordered agents to report on people with communist sympathies or German or Italian backgrounds, and President Franklin Roosevelt directed him to continue.[8]

Repression in America

Some of these moves were deliberate and official, such as the Palmer raids of the 1920s[9] and the McCarthy period of the 1940s and into the 1950s.[10]

But there was a constant refrain of anti-communism starting with the Bolshevik revolution of 1917.[11] The American Farm Bureau Federation, for example, labeled anyone involved in the Farmers Union or any cooperative movement a dangerous Bolshevik.[12] The Farmers Union responded by reversing its opposition to the Cold War and the Korean War and conducted its own purge.[13] Anticipating that they would be Red-baited, labor unions preemptively expelled communists from their ranks, an act that even anti-communists today say became a turning point in the power of unions in the American workforce.[14]

In 1938 the House Un-American Activities Committee (HUAC) was established with Martin Dies as chair, and the Dies committee began investigating communists. There were analogous legislative committees in many of the states.[15] From the time of Winston Churchill's famous speech in 1946, an iron curtain fell over the American mind. University and college administrators and professors muzzled themselves before any House or Senate committee could blacklist them.[16] For the next twenty years, their students learned timidity until they exploded into the streets in the sixties,[17] when "the Government of the United States seemed like nothing but a brutal machine run by obvious maniacs . . . and their cracked minions . . . determined to keep corrupt little tyrants around the world in power even if that meant the sacrifice of the lives of all of the young men of America."[18]

In a 1984 reprint of a 1934 Tess Slesinger novel, Alice Kessler-Harris and Paul Lauter wrote: "Within a few years, the entire world of left-wing culture was ploughed under by the attacks of McCarthyism. It is not always easy now, thirty years later, to perceive how successful the campaign was in silencing and burying writers on the left."[19]

Granville Hicks wrote in 1954 that his own personal life and other examples showed clearly that "teachers were constantly being told about the glories of freedom of thought and speech, but in practice they discovered that discretion was essential to professional advancement. Most teachers conformed"[20]

The same year Maxwell Geismar wrote:

> But now, in this new epoch of personal freedom for artistic expression, we
> have also moved into another period of social and political conformity—of
> timidity, fear, and suspicion. Ours is also an epoch of shifting social foundations
> in which the freedom, not only to dissent politically or socially, but even to
> criticize, has in turn become suspect. Those forces of self-appointed censorship

in our past, those forces of ignorance and superstition, have now gained a semiofficial status in our society and have gathered to themselves a semilegal sanction. . . .

For, if an artist has not yet felt the lash of official condemnation, he has already been sensitized to the less tangible but equally ponderous and oppressive forces of intellectual regularity in our present cultural atmosphere. Think for yourself—but think like everybody else.[21]

But by then a generation of writers had been disappeared, and Paul Corey found that he could think for himself but like everyone else in his do-it-yourself instructional articles and books on how to build things like houses and furniture. The relationships were mechanical, not social; they were demonstrable, not debatable. While the naturalist style of writing he developed for his farm novels was well suited to the DIY medium, his days of writing about Midwestern radicals and farmers were over.

The Crash and the Hope

What had caused the most vicious assault on liberty America citizens had ever seen? Many viewed the thirties as a time of hope and optimism.[22] The boom of the 1920s ended in the crash of 1929 and the Great Depression that set in by the next year. But there *was* hope. The Soviet Union, organized by its soviets, cooperatives of all the people who worked in a place, had found a new way in communism. From America the Soviets borrowed technology, modern organization, and the means for industrializing their agriculture. Many Americans went to the Soviet Union to participate.[23] If capitalism had failed, a new system was on the horizon. All it would take in the United States was a revolution of the proletariat. And the previous decade had provided promising signs for such a revolution.[24]

The Patterson Silk Strike of 1913 showed that working-class people were capable of organization; the 1914 militia massacre strikers at Ludlow, Colorado, showed the desperation of the capitalist class; the 1919 general strike in Seattle seemed to be the realization of the Industrial Workers of the World's vision of One Big Union and a revolution by general strike.[25] But the Palmer raids of 1920 decimated the IWW, sending many of its leaders to jail and deporting the rest with other radicals. So the proletariat was ready and primed for revolution, but officially sanctioned massacres and raids forestalled that by decapitating the leadership and intimidating the rest to foreclose that path to revolution.

Critic Granville Hicks wrote in 1933 that the answer to the doubts of the generation was to be found in an alliance of writers with the proletariat in its struggle for revolutionary change,[26] a new attitude in American literature as in American life.[27] The westward expansion had offered hope of independence and perhaps wealth that had fueled individualism, but as industrialization became dominant, the implications of capitalism became clearer and clearer. The lines of class struggle were so sharply drawn that no one could evade them.

Culturally, the new order was not sapping away an old established order; there *was* no indigenous American culture to undermine. Industrialization was creating a new civilization. The alternatives were struggle and flight. The public demanded interpretations of the chaos that surrounded them, and writers responded. Now the writer, "if he is to express whatever is vital and hopeful in the American spirit, must ally himself with the working class."[28] In the 1930s writers were exposing the forces behind the misery and destruction that all observed and many lived. Thus they revealed the conditions that made revolution inevitable.[29] Authors could either accept and defend the existing system or oppose it and work to overthrow it.[30]

A revolution required a well-organized vanguard to lead the proletariat to final victory as the Bolsheviks had done in Russia in 1917. Trotsky had outlined the conditions for revolution: the ruling class loses faith in itself because it cannot change conditions; there is a struggle among cliques; people place hopes in miracles and miracle workers. Malcolm Cowley wrote in 1933 that "these words of Trotsky's might be written yesterday or tomorrow by a Washington correspondent."[31] But then, as now, the proletariat was divided, "white against black, native against foreign-born, skilled against unskilled, and is disheartened by its past defeats."[32] While the Communist Party USA could be the revolutionary party, it was not, chiefly because the proletariat did not accept it. There will be a revolution, Cowley predicted, in five years or fifty, but it will follow the recent history of Germany, not Russia. It would be fascist, not communist.[33]

American intellectuals turned to the Communist Party in the 1930s.[34] In the early part of the decade, the socialists and communists of Germany had fought themselves to a stalemate and allowed the minority of fascists to gain power. In Spain, people watched and learned the lesson. A few years later, all of the left parties allied into a popular front to elect the government of the Republic. A popular front was announced in the United States as well. It would include liberals, socialists, and other left-leaning people and bring them into the orbit of the Communist Party. In 1935 at a convention of writers, the party organized the League of American Writers. Most were "fellow travelers" rather than members of the party.[35]

Those included poet Ruth Lechlitner and her husband, novelist Paul Corey, who lived on a homestead near Cold Spring on Hudson in upstate New York.[36] Though both had roots in the Midwest, they were now engaged in the New York literary world. Both were members[37] of the League of American Writers if not the Communist Party, and both signed the league's poll in support of the Loyalists in the Spanish Civil War.[38] But neither was very political nor did they ever have much to say about the Communist project.

Women were a large part of the party even though both its ideology and practice subordinated and opposed feminism.[39] Lechlitner was more active in the league than Corey. She chaired its Book Distribution and Organization Committees,[40] taught writing under the league's auspices, and served on its And Spain Sings Committee to collect poetry in support of the Loyalists to be published as a volume under that name.[41]

In 1934 editor Willard Maas offered Ruth three pages devoted to revolutionary poetry in a special issue of the short-lived literary journal *Alcestis* and offered to facilitate the publication of any book-length manuscript she might submit.[42] The same year, another editor, Philip Rahv, invited her to write a comment on revolutionary poetry for the *Partisan Review*.[43] In 1935 Paul Rosenfeld, who was editing W. W. Norton's annual anthology, *The New Caravan*, wrote Ruth accepting several poems, including "Lines for an Abortionist's Office."[44] In 1936 the league's Executive Committee asked Ruth if she would be able to serve on the committee as they wanted to nominate her.[45] However, neither she nor Paul belonged to the party.[46]

Un-American Activities

The optimism of the decade was crushed by the Soviet-German pact of 1939 and the Stalin trials.[47] The Dies committee, established in 1938 to ferret out communists, began the finale. The Subversive Activities Control Act of 1950 contained an Emergency Detention Act under which concentration camps were built in the United States for communists. Twelve of these were maintained until the repeal of the act in 1971. Director Hoover kept a list of citizens that were to be interred, though he was never instructed the round them up.[48]

There was no statute of limitation for being on the list of disloyal citizens. You were on it forever. This was not a blacklist, but it was a gray list. Writer Howard Fast said, "The FBI killed the social novel."[49] The FBI blatantly intimidated every major American publisher to prevent the publication of Howard Fast's *Spartacus* and even the republication of his previous works.[50]

In 1988 Paul Corey recalled that he concluded that he was on such a list when his agent advised him to explore the market for "how to" articles rather than the kind of fiction he had been writing, because the witch hunters were not looking there.[51] That communication could have come at a meeting in New York City or a publisher's cocktail party.[52] There were no official bans, Corey wrote, but publishers became wary of such writers, and a writer never knew exactly what was behind a gently worded letter of rejection.

The market can disappear a writer just as surely as a commissar. Such was the fate of Josephine Herbst, to name but one example.[53] For reasons of poverty and self-employment, she was immune from the blacklist, but she was on a gray list that constricted her opportunities.[54] Such writers were smothered into oblivion by constricting their opportunities. Or it may be, as Kenneth Bindas wrote, that these writers faced the dilemma of having to work within a system whose destruction they advocated; that there was a basic contradiction between "a heightened sense of class solidarity within the reality of a system that requires commodification of the same ideology."[55] The writer's task was to convert a critique of capitalism into a commodity for a corporate publishing industry to produce and sell in a mass market while the corporate system was rewriting American culture to suit its goals, which included making class invisible rather than promoting class struggle. Any literary product that was incapable of bridging that dilemma could not see print. Witness Fast's *Sparticus*, which he had to publish himself for it to see the light of day.

Even if a book could successfully bite the hand that fed its author, there were other sources of pressure on radical writers. John Steinbeck was too successful and well known a writer—too much a market success—for corporate resistance, but even he was subject to delays in getting passports, denials of official status during the Second World War, and other annoyances and harassments because of the opposition of the American Legion based on its negative response to *The Grapes of Wrath*.[56] Writers Josephine Herbst and Howard Fast were denied passports.[57]

With the repression of the Communist Party and left-oriented publications, Ruth lost the major outlets for her work. Though the slicks and *Poetry* continued to publish her work, the subject matter changed from the insurgent to introspective and became personal rather than social; when it was social, it was ecological and observational rather than radical—she perceived problems but ceased to advocate social solutions. By 1982, when Susan Ware surveyed women writers in the thirties, Ruth had become one of the "other writers of the decade."[58] When Charlotte Nekola and Paula Rabinowitz compiled their anthology of radical writers of the period, they included a single poem of Ruth's, and that from her *Tomorrow's Phoenix*, published in a limited edition as a collector's volume, though they do not mention that.[59]

Alexander Saxton was another example. He attended Harvard before working as a merchant seaman during the Second World War, worked in a railroad yard in Chicago, and moved to California to be a construction worker and carpenter. Harpers published his novel *Grand Crossing* in 1943, and in December of 1948 Appleton-Century Crofts published his *The Great Midland* about communist workers in the Chicago railroad yards. Immediately the publishing house renounced the 1948 book, and booksellers and reviewers shunned it so that its American sales were minimal.

Disappeared as an author, Saxton became first a carpenter and then a historian. HUAC summoned him to Washington in 1951 where he claimed the protections that the Bill of Rights affords Americans. That was his death knell as a writer. He lost his position as a writer for Paramount Studios and a teaching job at San Francisco State University. Afterward he published several pieces in magazines, but his experience with *The Great Midland*, which featured communists, caused his agent to suggest that if he wanted to publish his next novel, *Bright Web in the Darkness* (1959), he should excise any communist content. The FBI hounded his wife. When *The Great Midland* was republished in 1997, he wrote in the introduction:

> Was my inability to earn a living as a writer due to the drying up of my own aesthetic imagination or to blacklisting? Under such circumstances one vac-

illates between internalizing the fault and blindly projecting it. Both are bad medicine because the first erodes self-esteem while the other flames out in self-serving anger.[60]

Corey was in just that position.

In the 1930s there was a market for left-wing books, and publishers responded. Publishers were not victims of a communist conspiracy, "but of an illusion that proletarian novels would sell," wrote former communist editor Granville Hicks in 1954; he added, "Pro-Communist books would have appeared if there hadn't been a Communist editor in the whole United States."[61] But no editor who recommended the publication of books that failed to sell could long keep her job. "It is one thing," Hicks wrote, "to get a book published and another to get it read," as Paul Corey would learn.[62] Or, as Kessler-Harris and Lauter wrote of the feminist writings of the thirties:

> More profoundly, perhaps, the concern for working-class life and for under-standing the relationship between gender and politics was suppressed in the fifties as anti-communism, suburbanization and consumerism controlled the environment. As the cultural soil dried up, writers lost their audience. The run of stories, novels, poems, and songs that had established for an audience certain expectations, artistic conventions, and ideological perspectives, came to an end.[63]

Less overt was the great cultural revolution that had been underway since the turn of the twentieth century to normalize corporate rapacity. Anthropologist Dimitra Doukas discusses this in her ethnographic treatment of the Illion Valley in New York, and Elizabeth Fones-Wolf documented it more generally throughout the United States.[64] A long-lasting and contin-uous assault on a whole culture has continued for more than a century via the corporate-fueled conservative think-tanks and cultural institutions that shape news and perception throughout the United States. It has been largely successful.[65]

One part of this program was to aggressively attack the basic tenet of communism, the existence of a class system, much less a working class in opposition to an owning class.[66] Rather, any difficulties workers may feel are the fault of their own individual failings while the wealth of the owners is a reward of their individual merits. Neither poverty nor wealth is attribut-able to larger structural processes that could be changed by political means, as left-oriented writers had argued. Rather, both are consequences of natural forces as unalterable as the weather.[67] With the success of the corporate cultural revolution, there was no intellectual purchase for the psychological, cultural, and sociological underpinnings of left writers, much less for any political movement.

So the cultural revolution dried up the market, and a generation of writers were disappeared into the maw of the corporate revolution. With that came the enforced memory loss and reinterpretation of the thirties.[68] "The mission for future scholarship," wrote literary historian Alan Wald in 2012, "is to reverse forced forgetting."[69]

The Death of Naturalism

Some write that by 1961 literary naturalism in the United States was either dead or transformed into something else, but certainly had fallen into obscurity. Some date the end of naturalism twenty years before that.[70] Others argue that this gritty style of writing refused to die because it is intimately connected with hard times in the United States. First in the 1890s, then the 1930s, the late 1940s and early 1950s, the late 1960s with Vietnam and Watergate, and as of the early 1990s, it was still with us because of the enduring reality of a rhetoric of opportunity juxtaposed with the realities of poverty.[71] While it is attuned to American life, "in its concreteness and circumstantiality, it also runs counter to the predominant strain of optimistic moralism in the American character."[72] So it is an uncomfortable fit.

Farm fiction coincided with the first couple of waves of naturalism.[73] Corey self-consciously modeled the writing style of his farm novels on French novelist Emile Zola's "naturalism," in which the text would be impersonal, disinterested, objective, and based on observation and detailed description to show how characters' fates were governed by detached, deterministic forces rather than any individual choices. This is the inverse of the message of the corporate cultural revolution, which attributes success to individual merit and failure to individual shortcomings.

In the twentieth century Corey's naturalist farm fiction did not enjoy a high reputation. The public increasingly turned to urban life and values and could no longer take seriously novels of farm life. Readers relegated farm fiction to "the limbo of the sub-literary."[74] While the best of the farm writers might compare well with the urban naturalists, farm novels were better as illustrations of naturalism than as works of art: "The authors employ the techniques and reflect the attitudes of naturalism without displaying the moral imagination, esthetic judgment, and technical skill required to produce truly superior fiction."[75]

John Flanagan proposed that the mediocrity of Midwestern farm novels consigned them to oblivion. Their style was competent, accurate, simple, direct, but stereotyped and impoverished. Their descriptions of rural life were faithful and complete with perhaps an excess of detail. Narrative techniques were bare and unoriginal, and the plots lacked intensity. "Good storytellers," among whom he includes Paul Corey and his Mantz trilogy, "are rare among these novelists of rural life."[76] All these are criticisms that various editors showered on Corey throughout the thirties and forties. And oblivion was his fate, along with that of many other farm novelists and naturalist writers.

Roy W. Meyer concludes that Corey's farm fiction is the best that treated issues realistically and significantly but that this kind of novel is not distinguished by more than technical competence. Thus, while it may inform readers, it does not move them. "In no middle western farm novel written in English is there the combination of stylistic power and realistic treatment of significant material that makes for the very greatest of fiction."[77]

In Meyer's view, while Corey's fiction may represent the farm novel in its most highly developed form, his slight plot threads were submerged under the wealth of detail so that they lacked emotional intensity. They had less impact on readers than other more tightly knit works even if they remain the most important examples of farm fiction.[78] Furthermore, since most readers by that time were not familiar with farm life, they were "likely to be more bored than impressed" by the details.[79]

Perhaps because increasingly few people lived in rural areas and were familiar with life on farms; perhaps because of the less than exhilarating pace of life on farms; perhaps because of the lack of great writers, farm novels decreased in popularity after 1940 until they ceased to be a meaningful category of American fiction. Paul Corey, like the naturalist farm fiction he wrote, fell to the urbanization and mechanization processes he described in his novels. He was disappeared by the sheer weight of historical processes. It may be that the thirties and early forties represent the emphasis of labor and increased significance of working-class people in American culture, as

Michael Denning suggests, but only if we accept that the production of food does not entail labor. Denning does not discuss the literature of rural America, perhaps because it was simply not significant.[80]

John Steinbeck was subject to the same political pressures, but *The Grapes of Wrath* (1939) was a major publishing success.[81] While Corey was writing about a subject alien to the experience of most Americans, Steinbeck was writing about subjects close to most. Everyone had wrestled with a failing automobile, changed a flat tire, been driven from one town to another looking for work or a better place to live. Everyone had experienced the death of the older generation and the adversities of the younger. Farmers' associations and chambers of commerce might denounce it as inaccurate and profane, but there were waiting lists at bookstores and libraries.[82] In 1942 critic Maxwell Geismar wrote that *The Grapes of Wrath* "is a sociological catalyst; in a moment it has transformed and given meaning to the chemistry of a decade of social change." And, "Consider this popular novel filled with new and perplexing concepts: mass unemployment, hunger, rioting, armed reaction, and revolutionary urging."[83] In one way or another all ordinary Americans had experienced these tribulations.[84] Geismar wrote:

> Tens of thousands, and then hundreds of thousands of eager Americans thumbed a ride in the engrossing pages of Steinbeck's novel, took a literary hitch on the Joads's jalopy, struggled west along Highway 66, perhaps the most historic route in contemporary literature.[85]

Because Steinbeck successfully fused contradictory impulses into a kind of "stylistic popular front,"[86] there was room for everyone.

Social History

In 1984 the president of Pennsylvania's Lock Haven University wrote to Paul Corey to say that when he was a teenager in the 1950s in rural Ohio, he had read Corey's *Three Miles Square* and that it rang true. He had the university librarian locate a copy of the book for him and reread it with the perspective of time and then read Corey's *The Road Returns*. He was looking forward to reading the third of the trilogy, *County Seat*. He thanked Corey for writing books that were accurate portrayals of the lives of farm people and asked whether there was a way he could acquire the three books.[87]

Corey responded, "It's impossible to express the pleasure and satisfaction your letter of March 22 gave me. A reader has found in my books what I had hoped I was putting in them." He wrote that in the fourth book, *Acres of Antaeus*, he covered the years "when the small farmer was driven from the land by Agri-business." He continued:

> Of course I'm not the first writer to be forgotten, or feel slighted. I felt that I was writing social history, accurate in fact of incident, and in human behavior. The books were critically well received, but the publishers chose not to push them. That happens to many good books. Mine have been out-of-print for forty years.
>
> I have one copy of each book still. . . . I sort of keep these copies for my estate.
>
> A niece of mine has gone to considerable effort and expense to get copies of these books. That is probably the only way they can be acquired now.[88]

He concluded that he had let the copyrights lapse so they were "up for grabs, but no one's grabbing. . . . Personally, I feel there is no better social history of that period in the middlewest."[89]

2

THE PROPHECY AND ITS FULFILLMENT

"These pioneer realists, pioneer moderns, in truth, were . . . unprepared for
the onslaught of a capitalist order and its corresponding ethic . . . and they
foundered. In their foundering is half the story of their time and of our own."
  Alfred Kazin, On Native Grounds, *19*

The End

Those bastions of Jefferson's democracy, family farms, did not fade away
because of bad management or ineptitude of farmers. They were systematically destroyed. The process took a hundred years, but it was clearly visible
by the 1930s when Paul Corey started writing about it. It was a process of
government policy to set in place a structure that would favor large producers over small ones and that would force family farms out of existence
to industrialize agriculture. In the United States, capitalism achieved what
communism failed to achieve in the USSR, factories in the field and the
death of family farms. The institutional means for this accomplishment
included the Morrill Land Grant universities such as Iowa State University,
the extension service, the Farm Bureau, banks, implement dealers, and the
US Department of Agriculture. With the family farms went the cooperatives, the forms for economic democracy, not by inattention or corruption,
but by explicit policy.

DOI: 10.5876/9781646422081.c002

Today's food crisis with attendant epidemics of obesity and associated disorders is part of that harvest. Another part is the political corruption of agricultural states, among them Iowa. As anthropologist Walter Goldschmidt said in his classic *As You Sow* (1947), we are now harvesting the social maladies that are the products of those processes.

A Prophecy: The View from 1944

In his 1944 juvenile novel *The Red Tractor*, Corey deals with many of the issues he treated in *Acres of Antaeus*, but more consistently from the points of view of small farmers. One of the characters, Charley, says:

> Their big argument is that Fairview [a corporation] can produce more food
> on their big fields [5,000 acres] per acre than us fellows on our family-sized
> farms. They claim they're more efficient and when the country and the world
> is hollerin' for grub, we're standing in the way when we don't let 'em take
> us over. The worst of it is, they hold all the big cards—money, politicians,
> newspapers.[1]

Fairview, a corporation with the sheriff, implement dealer, banker, and county agent on its board of directors, has all of the tractors in the county, 5,000 acres of land, and workers from Chicago. It has offered to buy the Shieldses' 80-acre farm. These words are from a neighbor who has received a similar offer. Their response is to go in together to get a bank loan to each buy half of 160 acres from a third neighbor. Jim Sheilds comments on their recent acquisition, "It's not much of a farm." "That it ain't," agreed his neighbor [Charley], "but it could be built up. It's been cropped to death."[2]

> Charley's fingers worried the tablecloth. "You got to do your own figurin',
> Jim. But a family-sized farm is somethin' you can pass along. The homeplace
> ain't somethin' to get shed of lightly." His eyes appealed to Jim. "It's more'n
> just you folks. More'n fifty percent of Iowa farms is owned by people who
> don't farm 'em and the rest ain't full-owned by the farmers that are on 'em.
> That's goin' to ruin us all if we let it go on."
>   For just an instant Charley's pale eyes flashed around the ring of faces. "It's
> time us small farmers got our backs up. We gotta do somethin'. The Farm
> Bureau's dyin' out around here, and they never did give us farmers any say in

the runnin' of it. They couldn't be trusted to help us little fellows. We gotta find a way to help ourselves."[3]

Later, a father and son are waking:

Stan [Jim's son] and his father walked along in silence. They stopped a moment on the bridge over Indian Creek and leaned on the steel railing looking into the muddy water of the dredge-straightened stream below.

"There goes a lot of good Iowa soil," Jim said.[4]

They step over a dilapidated fence and walk through a field of spindly corn stalks. Corey continues, "Years of continuous cropping had broken the soil and rains had washed ditches and gulleys until yellow scars stood out on the valley side."[5]

The novel goes on to tell us that:

When Stan took the Future Farmers courses in the Hanley High School, he'd studied about erosion—sheet erosion and gulley washing. He used to dream, to pass the time plowing corn [with mules], of getting an old piece of worn-out land cheap and building it back into production. That'd be fun. He figured how he'd countour[sic]-farm it and strip-farm it, maybe even put in terraces. Then he'd put dams in the big gulleys and set out black locust trees to hold the soil.[6]

Mr. Slater, the county agent from Iowa State University, addressed the 4-H Club as part of a plan to help Fairfield take over the club:

He talked about the great need for food in the world—the millions of tons that must be produced to keep America and other countries from going hungry.

"But," he said, "throughout this state antagonism is growing up among the small farmers against the big farms and industrialized agriculture. This is a dangerous situation. These big organizations can produce so much more so much more efficiently than the smaller units that everything should be done to smooth the way for them. In many instances small farmers should welcome the absorbing of their farms by these organizations and work for them because that would mean a greater production of food to feed the starving world."

"My message to you young people," he finished, "is to do all you can to help this great trend toward more efficient farming along. I might add that it is your patriotic duty in the face of the great need for food in our country and the world to do this. Thank you."[7]

The young members of the 4-H Club elect the father of one of their members, a small farmer, Mr. Horton, to be director of the club. He addresses the members:

"The one thing you kids have always got to do is live up to your four Hs.[8] Never let 'em slip for one minute." He stroked his chin with a crooked finger, then went on, "Peggy [his daughter] told me about Mr. McKenna's [the agent from Iowa State University] message to you kids the other night. There's a good example of powerful forces in the country trying to use a fine club like yours for its own ends—this time it's Fairview. You must always be on your guard against that kind of trickery. Now, I've said my say—the job's yours from now on and I won't say anything more until you ask me."[9]

Four neighboring farmers are in debt to the bank for one reason or another and have to come up with cash to make payments by the end of the harvest or face losing their farms. They band together to approach the agent of the US Department of Agriculture's Farm Service Administration to get a government loan, and find a way to apply pressure on the implement dealer to make the next tractor available to them. Having bought the tractor, they can finish their farming tasks in time if they operate the tractor round the clock. But they still have to thresh their oats. Fairfield attempts all kinds of sabotage and has tied up all the threshing machines, so none is available. Then some youngsters chip in on a junked-out implement, tow it home, and rebuild to get the oats threshed and to market just before the debts come due.

This book, Corey's *Red Tractor* (1944), prophesized accurately what was to happen over the next seventy-five years. So did Theodore W. Schultz's *Agriculture in an Unstable Economy* (1945), which he prepared for the influential postwar planning initiative, Committee for Economic Development.[10] University of Chicago agricultural economist Schultz makes it abundantly clear that after the Second World War there would be too many farms operated too inefficiently by people too poor.[11] Schultz's major contribution was to see farms as parts of a larger economic system, interacting with it at every juncture. The excess labor would have to be absorbed by secondary and tertiary industries, service sector and government work. These sectors would need to grow much more rapidly than agricultural productivity.[12] Rural children should be educated to provide urban opportunities for them. He roundly condemned the back-to-the-land movement for promoting inefficient farm size and organization. Policy should discourage subsistence and low-production farms.[13] This semi-official advice is consistent with the country life movement's emphasis on increasing efficiency. This book was directed at policy makers.

On January 25, 1942, Corey wrote to the Department of Agriculture regarding its policy of encouraging chain farms, consolidations of smaller farms into larger units, to increase production for the war. Russell Smith, director of economic information, answered there had been no change in the department's longtime policy of conservative farming on units of optimum size to support individual families. He continued that disruptions in trade had mandated increased production of oil crops, among them soybeans, and in some cases shifting from extensive grain farming to livestock farming.

> It seems to me that the trend toward an optimum or adequate "family-sized" farm may well be emphasized. However, any tendency toward a decrease in the large-scale units probably will be more than offset by the abandonment of farms too small to support a family adequately. More and more of these farms will be abandoned as low-income farmers go into war industries or military service. The abandoned units are likely to be leased or purchased outright by remaining farmers.
>
> In the Western Corn Belt and the Eastern Great Plains, the most satisfactory family-sized farm would be considerably larger than the average or typical farm of recent years. The abandonment of farms in the Plains during the drought years made it possible for many of the remaining farmers to round out more economical operating units.
>
> In Ward County, North Dakota, the recommended minimum unit for a family-sized general farm is 480 acres. At last report more than half the general farms in the county were smaller than that.[14]

He went on to say that any far-reaching change in farm size would be a long-term process, not especially related to the war, though it (the war) might "facilitate some desirable adjustments previously blocked by the lack of alternative employment for low-income families."[15]

Corey wrote about farmers, including "low-income" farmers. He was not romantic a dozen years earlier in his depiction of squalid conditions on a small farm in upstate New York in his 1934 story, "Their Forefathers Were

Presidents," in *Story*. Eight dirty barefooted, hungry children play in the dirt behind the house while their mother cooks in a sagging lean-to. The father and a visitor drink in front of the house. "You'll get your head knocked off if you go around in front," says one sibling. When the men come to the back of the house to eat, the father announces, "This is on the county." Finished, they return to the front with a bottle of rye while the children fight over the scraps.

A brother and sister walk until they find a woodchuck with its foot caught in the jaws of a metal trap and graphically club it to death. The father lies in the front yard in a drunken stupor while the guest has sex with the mother in the bedroom in the presence of a sleeping infant. The triumphant children return to the house with their woodchuck.[16]

The next year, in the same magazine, Corey wrote about a Danish immigrant hired man courting an American farm girl, the support he gets from his boss, the tormenting of the other young men, and finally the noisy charivari at the newlywed's house. He concludes, "But love wasn't always like that in the middle-west, two decades ago, even though life was smooth for the country folk; and it's never like that anymore."[17]

Through Greenest Iowa—the View from 2021

Corey told the story of the death of Iowa's farms in ethnographic detail. How does it look from the perspective of nearly a century?

Interstate 80 enters Iowa from Illinois on a bridge. In the Mississippi River below are barges three abreast and fifteen long loaded with Midwestern corn going south to the Gulf of Mexico. The highway moves straight west into Iowa between fields of corn planted so densely it can't wave. A sea of green in all directions. There is an occasional shimmering low building, sometimes several in rows—the confined animal feeding operations or CAFOs full of pigs fattening for their trip to the meat-packing plant. This is Iowa's most visible and famous industry: agriculture. When there were farmers, it was called farming. Now it's called an industry.

Off the interstate, on the county roads, the corn crowds in more closely. The West Liberty exit puts drivers on the county road X-40, which goes south to the town where in 2011 the high school graduated its first class bilingual in English and Spanish. At one end of the town is a massive turkey processing plant. At the other are several Mexican restaurants. In between is a Chinese place, a Mexican grocery store and bakery, a restored movie theater, a few dark storefronts, and a feed store. To the north of the interstate, X-40 goes to Springdale, the Quaker village that was a stop on the Underground Railroad where John Brown hid out before his raid on Harper's Ferry. By then, the roots of the industry were already in place—railroads, grain elevators, Iowa State University.

The fences and any trees have been removed to expand cultivation up to the ditches along the road. Some smaller county roads have been plowed under. Always expand production. That's what industry does. Any waterways have been smoothed out, replaced by underground pipes—or tiles, as they're called—to collect excess water and carry it into the nearest watercourse and on into the Mississippi, where it will join the other poisoned water on its way south to kill life in the Gulf of Mexico.

More or less evenly spaced along the roadside ditches are overgrown driveways. Those are the remnants of farms, places where families once lived and wrested or coaxed their livings from the ground. There may be a decrepit house surrounded by a thicket of trees and brush. That's a place scheduled to be bulldozed to make room for corn. Such was the fate of the house where Paul Corey was born—the house his father built.[18] These days, when people want to sell their land, a house, no matter how well maintained, detracts from the value of the land because of the cost of demolishing it.

Today there are no farms in Iowa. There are operations. There are places people call farms, but their owners live in Chicago, Los Angeles, or New York. A farm management firm sends the check. That firm rents the land to an "operator" who may work 10,000 acres with machines as big as the houses that were torn down to make way for the operation.

Anything that resembles the iconic red barns that used to dot the landscape to mark each farm is weathered gray, its roof sagging. It may be full of forgotten hay rotting as the loft drops closer and closer to the ground with each passing season. Those barns were built to house and feed cattle, horses, and mules through Iowa's long winters. Now the only livestock is in biosecure CAFOs. And the operators all have machine sheds.

I-80 soon comes to two exits for Iowa City, either of which leads to the University of Iowa with the golden dome of its Old Territorial Capitol glistening in the sun. It was there when Paul Corey came in 1921 and when Ruth Lechlitner came in 1925.

Farther west are exits for the Amana Colonies, those communistic Ana-baptist religious settlements that transformed themselves into a major cor-poration in the 1930s when most corporations in the United States were going broke. Then come exits for Brooklyn and What Cheer, where Frank Luther Mott was a newspaper man before he joined John T. Frederick to edit *The Midland* in Iowa City and to found a school of journalism. More or less in the center of the state is Des Moines with its own golden-domed state capitol building. There I-80 intersects with I-35 running north and south.

Everything in Iowa is North, South, East, or West, laid out with the flawless logic of Thomas Jefferson's grid. Iowa was part of that great exper-iment in social engineering to provide the economic underpinnings for democracy, family farms. A square mile is a section. In the middle of each section there would be a school to educate the children of the surrounding farmers to the values and principles of democracy. It was in these schools that the children of the German, Yankee, Danish, Norwegian, Polish, Czech, Welsh, Irish, English, Slovak, and Dutch farmers fought together and learned to get along together, where they forged English into their own dialect.[19]

That's the dialect of radio and television because it has become standard for American media. It is average. If it's any time near an election, hopefuls introduce themselves with the standard Iowa candidate's opening, "I grew up on a farm that's been in our family for a hundred and fifty years." They may be still living on the rent from that land.

North on I-35 is Ames, where Iowa State University is. Iowa State was a big part of the transformation from the farms Paul Corey wrote about into today's industry. Farther north the CAFOs get more numerous and closer together to the Minnesota border. South from Des Moines is Missouri. Farther west on I-80 is an exit for Marne, close to the farm where Paul Corey was born. Past that exit the highway dips south to Council Bluffs before it crosses the Missouri River into Nebraska. North is Sioux City, gateway to South Dakota. It was here that the Farm Holiday Association blockaded the town in 1932.

The Dollar Economy

Paul Corey started working on the first of his farm trilogy early in the 1930s.[20] In 1964 he wrote that he meant his trilogy to be a social history of farming before the battle between family farms and corporate farms broke out in the mid-1930s.[21] But that wasn't the real battle, though Corey could not have known it. That could be handled by legislation, first passed in 1975, to limit corporate ownership of farms. The real enemy of the yeomen farmers of Jefferson's grid was the industrialization, not creeping but thundering in with the railroads and telegraph lines to incorporate western lands and agriculture into the eastern industrial system. The railroads carried goods and people; the telegraph brought market news.[22] More poetically Robert West Howard wrote, "Dollar Economy rode like a thousand-tailed comet across the purple sky" and mined America's soils.[23] That process was well under way by the time Corey was born early in the 1900s.

Today the back roads through Iowa's small rural towns are lined with dark storefronts, a bar or two, maybe an agricultural implement dealer with brightly shining machines on display, an early twentieth-century stone or brick building that used to be a bank, now standing empty, sometimes a grocery store or local restaurant where old people linger over the daily special and a piece of pie. In the mornings old men sit around a table to gossip and drink coffee. Sitting around another table are the old women. Flanked by a field of headstones in orderly rows there is a substantial church towering above all. And by the railroad tracks, or where they used to run, perhaps now a bicycling path, is a grain elevator.

During the last half of the nineteenth century grain elevators popped up across Iowa's landscape like mushrooms after a rain. They were part of the infrastructure for the high-volume sales that integrated Iowa farms into an international system of commodities—buying, selling, shipping, and processing. Today many of them have concrete pads where semi trailers dump bumper crops of corn to await transportation, load after load, until there is a golden cone of corn.

The patriarch of an old and successful community banking family commented that the dead banks of Iowa's small towns were gone because the bankers had children who were more interested in money than in maintaining community institutions.[24] But in Corey's trilogy the local bank is thriving, and its banker, Rhomer, while he may not be interested in the fate of the community, is interested in making money, not in selling out. He holds mortgages on everyone's farms and businesses. He knows the hopes and troubles of everyone in the county.

In *The Road Returns* the community-minded farmer named Crosby organizes a cooperative grain elevator, and Rhomer, instead of lending the cooperative $6,000 when Crosby cannot sell all the shares to farmers, buys 60
of the 160 shares. Thus, he controls sixty votes. Crosby points out that all
Rhomer has to do is get twenty farmers to side with him and he can control
the elevator. Crosby's friends don't think Rhomer can do that. Later, when
farmers come to the banker for loans, rather than taking land or equipment
as collateral, Rhomer offers to buy their shares of the cooperative at sixty to
eighty cents on the dollar: "Rubbing his hands together [Rhomer thinks] just
a few more like this farmer and the Rhomer Bank would have a controlling
interest in the Co-operative Grain Company and that would be the end of
Ed Crosby's little scheme."[25]

Crosby recognized that it was a violation of cooperative principles to
allow one vote per share rather than one vote per member, but he needed
to raise the money to buy the elevator to put it into the hands of the farmers to manage. And then, as farmers got hard up and sold their shares for as
little as fifty cents on the dollar, the bank came to own seventy-five of the
hundred shares. When the bank went broke, the cooperative grain elevator
went, too.

Businessmen and the Failure of Markets

Today's farm operators are quick to say they are businessmen with their eyes
glued firmly to their bottom line. And so they are. That bottom line can be
hard to understand unless you take into account the host of subsidies and
protections from the Department of Agriculture. "Farming the government,"
they call it. They are thoroughly modern in their approach to business and
technology. Their tractors are guided by computer-processed information
from satellites to plant optimal amounts of seed and apply optimal amounts
of fertilizer to the fields they rent to expand production.

But the settlers who came to Iowa in the mid-nineteenth century did
not think of themselves as businessmen. They might be opportunists. They
might hope to gain a piece of land cheap, maybe build it up a little, sell it for
a lot more than they paid, and keep on moving west. But not businessmen.
Anthropologist Horace Miner, who studied farming in Iowa about the time
Corey's trilogy was coming out, noticed this element of land speculation.[26]

When the price of commodities rose after the First World War, land values also increased.

By 1920 land that had been worth $100 an acre was selling for $400. Many farmers mortgaged their land to buy more land, also mortgaged. "Farms were sold and resold, the price rising and another mortgage being added with each transaction."[27] With the Depression and the collapse of commodity prices, farmers couldn't meet their obligations on debts. Every year the Mantz family of Corey's trilogy worried about meeting their mortgage payments, if not for land, for equipment. They lived under the constant and recurrent threat of repossession. In fact there was a wave of foreclosures that provided Corey with the basis for his novel *Acres of Antaeus*. People who had not speculated on land and saved their money could pay their taxes during the Depression. The others lost their farms to tax sales or to insurance or mortgage companies.

Miner identifies two opposing philosophies of life among the farmers in the county he studied. Those of Yankee stock followed the risky speculative expansive practice. German and Norwegian settlers "were more inclined to value their immediate possessions and ways of life over the speculative advantages of future 'advancement.'"[28] They would not risk their farms to buy new ones for their children or to expand their production. They did not share the "unbounded drive for profit" of their Yankee neighbors but drew satisfaction from the present, even at the cost of being considered backward.[29] The sales of land and land turnovers correlated highly with the Yankee population except in one township settled by Quakers who held onto their land.[30] Many farmers were unable to purchase more land so the mortgage and insurance companies sold farms to city dwellers who then rented them to farmers. New Deal programs with their infusions of cash to farmers stopped the evictions. Otherwise, the whole ownership and tenancy pattern would have changed to one of absentee owners and tenant farmers.[31] Which is what it predominantly is today.

As the novelist of that time, Alexander Saxton, put into the mouth of one of his characters:

> More than half the farmers are tenant farmers aren't they?—they don't own their own land; and a lot of them are mortgaged up to the eyebrows, so they don't really own anything at all. . . . They're a rural working class, aren't they?

They don't own anything but their ability to work, and they're at the mercy of the landlords and middlemen. Most of them don't know it yet, but they'll find out pretty soon.[32]

The 1938 guide to Iowa compiled by the Works Progress Administration's (WPA) Federal Writers' Project reflects these developments:

February is moving time for the tenant farmers who must be ready for spring planting on their new farms by March. Trucks and wagons crowded with furniture rumble over the frozen roads and tired herds of cattle and sheep trudge in their wake. Load after load jolt past the farm houses, stretching out like a gypsy caravan going on to camp elsewhere for a brief time . . . [ellipsis original] hoping that the new farm will be a little better than the last one. Here is a tragic epitome of the problem of farm tenancy which is increasing so rapidly that there is real cause for alarm.[33]

The Roosevelt administration responded in 1933 with the Agricultural Adjustment Administration (AAA) to increase and stabilize farm income and conserve the soil and, with the Farm Credit Administration, to increase security of ownership.[34]

From 1933 to 1938 various government agencies loaned more than $4.5 million (equivalent to $89.5 million in 2019) to Hardin County farmers, where Miner was, mostly to avoid foreclosure. Those agencies *gave* farmers another $2 million (about $40 million in 2019). The system of price controls was meant to allow farmers to get enough income to pay off their new government loans. Miner concludes that without such intervention, the ownership pattern of farmland would have radically changed, as Corey foresaw in *Acres of Antaeus*. Written by the Farm Bureau, the programs secured the larger landowners in their ownership, the ones who had bet everything on expansion.[35] The AAA introduced programs of soil conservation, crop reduction, and government purchase and storage of corn to remove it from the market, called "corn sealing." But, Miner observed, "farm [cultural] values were not attuned to crop reductions."[36]

The whole farm philosophy since the settlement of Iowa, had been one of production. Every available means of increasing production had been eagerly grasped. A good deal of the prestige of agricultural departments and colleges was due to their contributions to this end.[37]

Government loan agencies urged farmers to produce; but other programs sought to reduce production. Farmers thought the market was at fault for the existence of a surplus, not the farmers for producing it. And farmers increased their production by the increasing use of hybrid corn.[38]

Ninety-five percent of the farmers entered the programs to secure the cash payments for crop reductions. From their point of view, this was recompense for soil conservation practices rather than relief. Farmers rejected the thought that the payments were to pay the difference between what the market offered and what the farmer needed. The farmers systematically misunderstood the goals of the programs, but they participated because the programs provided much-needed cash in a way they could rationalize to themselves and others. "Sealing" the corn by purchasing and storing it to take it off the market made more sense to farmers because it did not demand reduction in production, though the idea of buying dear to sell cheap the next crop season just confirmed the farmers' view that the government was stupid.[39]

The extension service and the Farm Bureau were central in the local administration of these programs.[40] In spite of farmers' opposition to being told how to farm, they ". . . grasped at any government plan which offered salvation from their economic difficulties."[41] In central Iowa the reception of the New Deal, with the exception of the farm programs, was negative, and Hardin County voters rewarded the Roosevelt administration that had saved their farms by voting Republican in 1940.[42]

Iowa's Henry A. Wallace, secretary of agriculture, was aware of these issues. In his 1933 report to the president, he wrote:

Some farmers. . . . believe we got into the present economic jam partly as a result of technical efficiency. They ask why Government agencies help farmers to grow two blades of grass where one grew before and simultaneously urge them to cut down their production. They declare it is almost criminally negligent for a Government to promote an increase of production, without facing the results of that increase. These ideas lead to something of a revolt against science, and to the demands for a halt in technical progress until consumption catches up with production.

It is undeniable that science creates problems; but the remedy is not less but more of the disturbing ferment. What we need is not less science in production, but more science in distribution, and this means distribution of wealth as well as of the physical products.

Science has magnificently enabled mankind to conquer the problem of producing enough to go around. It has now to help us utilize the increased productivity.[43]

Wallace's conception of economics is quite different from the contemporary notion. His understanding is much more aligned with what anthropologists today would call "political economy," or "political ecology," including many factors beyond the working of the market itself. He continued:

This is the special province of economics. It is a difficult field, because the data include facts of psychology, of politics, of history, of race, and even of religion, as well as of production and demand. Reducing such diverse facts to order is harder than discovering relationships among chemical elements insolated in a test tube. The economist cannot fix his material; he must deal with the living, changing, dynamic world. But the difficulty does not excuse evading the problem. It cannot be evaded.[44]

Miner observed that small-town businessmen thought of farmers as being dependent on them and minimized their own dependence on farmers. Thus they were antagonistic to programs that increased the income of farmers and their dependence on government agencies, even though most of that income flowed into merchants in the rural towns. Today the dead eyes of abandoned small-town businesses bear mute witness to the wrongness of this point of view.

Miner's final conclusion:

Crop restriction conflicts with the basic rural philosophy of maximum production. Nor is this philosophy just an economic one. It pervades the rural evaluations of men's status and even their character. In such a culture, payments to farmers to restrict their productive acreages will not change their philosophy of life.[45]

A student of Robert Redfield's at the University of Chicago, Miner saw the culture of Hardin County as an integrated whole rather than as a something shaped by evolving economic or political structures or policies. But Miner's conclusions were not to the liking of the Department of Agriculture.

Structuring an Industry

The first stage of industrial development in the East of the United States from the 1870s to the 1930s saw relatively slow growth of secondary organizations to coordinate and communicate among primary organizations. These secondary organizations included trade associations such as the National Manufacturers Association, the Chambers of Commerce, industry associations, and federations of allied labor unions. In agriculture, in contrast, these secondary organizations came first and provided the structure to guide the subsequent industrialization of agriculture. Among the secondary institutions were the Farm Bureau, (supplanting the grassroots voice of farmers—the Iowa Farmers Union, chiefly by Red-baiting), universities, experiment stations, extension services, the Soil Conservation Service, the Farm Credit Administration, all of the various commodity groups, crop and livestock statistical reporting services, market news, weather reporting services and the Rural Electrification Administration.[46] Within these structures, larger producers could flourish at the expense of smaller ones, just as in Corey's story that opened this chapter.

The findings of the 1938 Federal Writers Program's guide to Iowa are worth quoting at length because they were as prophetic as Paul Corey's or Alexander Saxton's.

> The modern Iowa farmer . . . [is] a progressive, business-like producer who sells almost 90 percent of his product and studies politics, weather reports, improved methods of production and marketing, crop control and soil conservation.
>
> An integrated structure has been built to aid him and further his interests. The Iowa State Department of Agriculture and the office of secretary of agriculture were created by the Fortieth General Assembly in 1923. The work of the department is subdivided. . . . Five associations are affiliated with the department—the State Dairy Association, Horticultural Society, Horse and Mule Breeders' Association, Beef Producers' Association, and Corn and Small Grain Growers' Association.
>
> Closely co-operating with the State departments of education and agriculture is the State College of Agriculture and Mechanic Arts at Ames. The extension service at the college works with the United Stated Department of Agriculture, from which it receives part of its financial support. The informa-

tion and other benefits provided by the experiment station, also connected with the college, are available to all. . . .

There is a county farm bureau in every county in the State. Organized in 1918, the membership of the State Farm Bureau Federation in 1938 was estimated at 50,000 families, each of which pays $10 [about $281 in 2019] in annual dues. The 4-H Clubs for boys and girls, associated with the farm bureaus, carry on educational work. County extension agents, employed by the county farm bureau groups with salaries set by them, direct the work of extending agricultural information. The United States Department of Agriculture and the extension service add contributions to the county appro-priations and the farm bureau funds for this purpose. A close-knit organiza-tion has been developed, starting with the county agents, and widening out through the farm bureau groups, the State extension service, the college at Ames, to the United States Department of Agriculture. Some opposition to this benevolent hierarchy has found expression in the Farmer's Union . . . and the Farm Holiday Association, the most important independent farm organizations in Iowa.[47]

It was this superstructure more than any other factor that was responsible for the death of family farms in Iowa.

Killing Cooperatives

After the setbacks of the nineteenth century, the cooperative movement with its populist ideology and organization reasserted itself and plays a prominent role in Corey's writing, not so much in the Mantz trilogy as in his juvenile fiction, in which cooperatives were the solution to controlling monopolistic practices whether in the Hudson's shad fishery or New York's dairy industry.

The program of the cooperative movement was to organize enough small producers of any commodity to control the market. The Farmers Union, or Farmers Educational and Cooperative Union, aimed to eliminate all capi-talist operations between production and consumption. It would do this by extending cooperatives to control agricultural commodities in production, transportation, processing, and final distribution to urban buyers. The rheto-ric of the union was anti-capitalist. Cooperativism would destroy capitalism, at least in rural areas. The language was neither socialist nor revolutionary but Biblical and Christian. They would "feed the multitudes."[48]

When this resurgence of populism began to show signs of success, railroad companies, the Chicago Board of Trade, banks, and chambers of commerce banded together to form an opposition organization, the Farm Bureau, to mount propaganda campaigns among farmers and organize alternative "safe" cooperatives. It would take the corporate cultural revolution to the countryside. By 1919 the Farm Bureau was a national organization.[49]

The Smith-Lever Act of 1914 allowed the Farm Bureau to graft itself onto the newly emerging extension services of the state agricultural colleges, and together they spread a doctrine of farmer as businessman. It was safe and respectable; it had none of the dangerous connotations of radicalism or Bolshevism in the period of the Palmer Red Scare.[50] Rather than the elimination of capitalism, there was talk of its encouragement and enhancement.

This was the official and institutional context for the development of agricultural science. A 1911 report of a committee of the Association of American Agricultural Colleges and Experiment Stations stated, "The term farm management may properly be restricted to that phase of rural economics which deals with the business organization and direction of individual farm enterprises."[51] The same year a committee of the American Farm Management Association, a component of the agricultural infrastructure, recommended a series of investigations to lead to "a safe working basis for the organization of the business of farming," and a 1913 book on farm management defines it as "the study of business principles in farming."[52]

The view of farms as rural business enterprises dominated the study of farm organization. Farm studies used categories of business analysis—capital, rent, and labor costs—and attempted to compute farm profits for different sorts of farm organizations.[53] One investigator reported that, "from a business standpoint a farm cannot be called successful that does not pay operating expenses, a current mortgage rate of interest on capital and a fair return as pay for the farmer's labor and management."[54] Others pointed out conceptual problems with this mode of calculation. C. C. Taylor and E. G. Hurd, for instance, wrote:

> The [family] farmer . . . supplies his own working capital . . . Contrasted
> to this type of business are many businesses which may be called capitalis-
> tic, in which nearly all of the cost goods are hired by the entrepreneurship
> instead of being furnished by it. In any business the expense of producing

the product, based on opportunity cost, is merely imputing to the factors which produced the product an amount equal to the prevailing market rate. In a capitalist enterprise, wealth is actually distributed on this basis. However, to employ this same procedure in a family enterprise involves considerable fiction. The total long time family income is quite largely a unit sum which for any particular year cannot be split into its elements on the accounting basis. However . . . we have been compelled to use the concept of profits as an index of success.[55]

Such accounting methods indicated that nearly half of the farms in three Iowa counties not only made nothing for the farmer's labor but failed to make 5 percent interest on capital in addition to operating expenses.[56]

Agricultural scientists created the dominant paradigms of agriculture and agricultural economics, and thereby defined the realities of rural America. To develop or improve agriculture would require research and development that, as in industry, would lead to increased productivity and larger units of production. This was the program of the agricultural colleges and the source of advice offered to farmers. Not only did it come from experts, but farmers could see themselves as the American ideal, businessmen, instead of the lowest order of the old world from which many had come, peasants. In this period, as in others, it was safer to be a capitalist than a radical. The respectable Farm Bureau, with its alliance with the experts of the agricultural colleges, had nothing good to say of radicals.[57]

Whether farms were capitalist enterprises was not an askable question. They should be. Agricultural agents should help poor businessmen improve themselves as the urban-based country life movement had long advocated.[58] The question was not how farms were operated or how to develop them in those terms, but how to achieve maximum efficiency by some measure.[59] The Farmers Union and other populist organizations argued that improvements in productivity would be of little value to farmers unless there were changes in the economic system so that producers could receive reasonable returns for their labor and capital.

Toward the end of the Second World War agricultural historian Robert West Howard would promote cooperatives as the wave of the future to insure economic democracy in America. A strong farmer cooperative movement would decentralize government as well as industrial power, be the death knell of cities, and herald a return of most of the people to villages and farms. No American city, he wrote, was a fitting residence for people. "So the path of America's future leads back to the land."[60] This was clearly Paul Corey's view in *Five Acre Hill* and *Buy and Acre*. Neither Howard nor Corey foresaw the development of the post war cold war and then the continuous warfare state in the United States that followed to bolster industry and continue the patterns developed during the war.

A Long Time Ago in a Galaxy Far, Far Away

Taylor and Hurd did not say just why they were compelled to use the concept of profit as a measure of success rather than developing one appropriate to the data they had, one based on their observation that the family's income was a unit sum that could not be further divided into components such as returns to capital, labor, management, and land. A world away another agricultural economist was grappling with similar issues in the Soviet Union. A. V. Chayanov suggested a system of accounting based on what he had observed within such households—the balance of household's need for sustenance and other requirements and the people's ability to satisfy it.

He concluded that farming households had a definite ceiling to their production, that it was not infinitely expandable beyond their needs. That ceiling, he argued, was determined by the balance of the family's needs and abilities.[61] That, he argued, was the difference between the economic logic of households on the one hand and firms on the other. Firms produced for the demand of an infinite market and had no production limits. In the United States, the Department of Agriculture and Iowa State were intent on converting households into firms.

But Chayanov's understanding meant peasants could not produce crops to meet the needs of others, especially in the cities. He suggested that the government could secure increased production by working with farming households to organize them into cooperatives so they could share in the product of their work and reorient their production goals. Meanwhile, those farming households that had more working members than consuming ones could get a little bit ahead for a while. But Marxist theorists from the cities saw in these better-off households incipient capitalism with all of its ills and wanted to excise them from the countryside.

Stalin had plans to industrialize Soviet agriculture by leap-frogging the reactionary peasants and establishing new industrial farms in Siberia, whose climate and agricultural potential are similar to Iowa's. Because Chayanov's policy conclusions did not match this program, Stalin had him shot in 1931 and put an end to any thought of farmer cooperatives as definitively as the Farm Bureau did in the United States. Because agricultural scientists from Iowa State University had worked out concrete plans for what industrialized farming would be, Stalin hired them to help him establish his industrial farms.[62]

The problem with the American agricultural economists' model was that it represented their dream of what they wanted farmers to be rather than the reality of what they were. But it was a dream that the Farm Bureau and Iowa State shared and that many farmers shared as well. As the exoskeleton of industrialized farming was put into place, farmers strove to be the businessmen that the extension agents wanted them to be. They expanded production. They went into debt to buy more land and bigger machines. And most of them lost their farms. It may have taken longer, but this program achieved Stalin's dream.

The agenda of the agricultural colleges was successful in multiplying cheap supplies of agricultural commodities for urban consumers, just as Stalin wished his former peasants would do. The American plan was not successful in insuring reasonable returns to farmers.[63] It never addressed the question of changing the economic institutions that governed farming in America.[64] To do so would have been to question the capitalist system as the cooperative movement did. Nor did it ever seriously consider the problem of preserving family farms. That wasn't part of the program. In fact, that was reactionary in a milieu of industrial expansion. Hence the opposition of the Farm Bureau to the Farmers Union.

The electoral and political successes of nineteenth-century populism evoked an anti-populist official and corporate response that strengthened in the face of such populist manifestations as the Farmers Union.[65] In alliance with the fledgling agricultural colleges, these interests redefined family farming as a business and developed business-oriented policies to guide farming. Faced with the Farm Bureau's visions of radicals, Bolsheviks, and revolutionaries and rejecting the old-country self-definition as peasants, farmers defined themselves as businessmen and followed the advice of agricultural experts. With each farm crisis the family farm became more tenuous until it has now become a nostalgic image. The dreams of full-scale industrial agriculture have been realized, leaving only remnants of dead and dying rural towns, collapsing barns, and driveways to nowhere.

After its heyday in the 1930s, the Iowa Farmers Union would receive its decisive coup de grace in 1953 when it was excised from the National Farmers Union during an episode of Red-baiting because its president refused to endorse Truman's cold war policy and his war in Korea.[66] The demise of the Iowa Farmers Union signaled the downfall of its cooperatives.

3

FROM IOWA CITY TO NEW YORK CITY

... all village life in the Middle West now seemed a cesspool of bigotry and corruption and even the very incarnation of joylessness.

Alfred Kazin, On Native Grounds, *193*

Publishing in the 1920s

Mass production, the machine age, and mass marketing created the newspaper and publishing industries and the professions of journalism and writing to supply them. For the first time in America writing was a profession, not something to do after hours of working for a wage in the customshouse or for an encyclopedia.[1] The center was New York with its publishing houses, literary agents, fellow authors, and jobs. There was nothing for Corey and Lechlitner in Iowa, and New York drew them like a magnet.[2] On the tail end of this first generation of American writers that could contemplate being professionals,[3] Paul Corey had written two novels, neither of which had found a publisher. But he was determined to join that profession. By the early thirties, Corey decided to write about what he knew best, rural Iowa at the turn of the century when he was growing up. And he was committed to writing.

The first sign that Paul Corey was headed out of Iowa was his choice of university and field of study. At Iowa State he could have stayed closer

DOI: 10.5876/9781646422081.c003

to home, learned farm management, horticulture, and all the allied fields to fuel the industrialization of agriculture in America. That he went to the University of Iowa to study the new field of journalism indicates not only that he wanted to be a writer but also that he hoped to join the stream of content providers the mass circulation publishers, newspapers, and magazines demanded and that journalism schools were beginning to deliver.[4]

To Chicago and Beyond

Paul Corey did not always write in a naturalist style. He had the manuscript of his first novel under his arm when he left Iowa City in June 1925 and went to Chicago, where he got a job in a bookstore that, it transpired, didn't hire "Bolsheviks" like him. He lasted three weeks. Next he worked as a reporter for a real estate magazine.[5] In July he received a letter from a Viking Press editor, George Ottenheimer, who was visiting Chicago:

> I read about 50 to 100 pages—enough to convince me that you can write and write well. Also enough—sad as it may seem—to justify my saying that your book has one chance in a thousand of ever seeing the light of day. It's much too poetic on theme and too aesthetic in conception to have any prayer of a sale. I found parts of it very immature—others quite sophisticated. I'd like to see it again when it's finally completed. I'm almost certain, however, that your chances are nil. Try something else and use your worthwhile talent in something more fitting your style. You could do fantasy admirably. I don't believe your method will ever fit a college novel. It's original—true—but it's very individuality and remoteness, it fails entirely to carry out a vivid picture.
>
> This novel compressed into a short story might be excellent. As 250 pages of text it grows too monotonous.
>
> I'm sorry to be so frank but there's no use misleading you with praise where praise is not due. Of course I may be, and often am, dead wrong.[6]

This was not his first rejection. Since 1920, magazine publishers had rejected nine of his poems and stories. In November, Ottenheimer wrote him from New York:

> Thank you for your kind letter of October 28th. I would be very much interested in seeing your novel when it is finally finished, and will certainly read it through.

If you come to town be sure to look me up.

With kindest regards.[7]

When Paul announced he was going to quit his job, he was offered a raise to stay, but he was determined to be a writer.[8] So he returned to Iowa City early in 1926 to complete his novel before taking the train east to Greenwich Village to find a publisher and take up the life of an author.

The Village

In 1915 Edgar Lee Masters's *Spoon River Anthology* challenged the sacred purity of America's villages. Its ruthless denunciation of America's scandal-ridden communities signaled what in 1922 Carl Van Doren called the "revolt against the village."[9] Anthropologists and sociologists had not caught up and would not for decades. They perpetuated the myth of the holy countryside, the purity of the rural, and the decadence of the urban. The literary repudiation of the holiness of the rural continued when Sherwood Anderson wrote that Europeans had managed to claim only sordid chaos from rural America's vast natural wealth.[10] And Sinclair Lewis of Sauk Centre, Minnesota, assessed the villages of America as militantly dull with the contentment of the quiet dead.[11] Young writers were fleeing this decadent lifelessness for the vivacity of New York City, center of the modern.[12]

> . . . . They had no desire to erect artistic monuments as such. They had emerged from the farms, the village seminaries, newspaper desks, with a fierce desire to assert their freedom and to describe the life they knew, and they wrote with the brisk or careless competence . . . that was necessary to their exploration of the national scene.[13]

In 1926 the streets of the Greenwich Village teemed with transient writers living in tenement basements, penthouse storerooms, corners of lofts, anywhere they could put up inexpensively.[14] Like Paul Corey, they didn't have anything special to say, but they knew they wanted to be writers.[15]

> There were many writers of talent, but few of professional seriousness and trained competence. All that has been changed in the last twenty years [since 1914]. Perhaps the greatest difference is in the number of writers who, by permanent standards, are second-rate and yet are intelligent and determined to

do their best work. Although they will never produce great books, they help to produce them, by creating the necessary background and the tradition that may nourish greater writers in the future.[16]

When Corey arrived in New York, new publishers were competing for new authors, and there were little magazines where a writer could get some attention if not any money.[17] Examples were the regionalist *The Midland* from Iowa and Jack Conroy's *Anvil* from Missouri that never paid contributors. But in the modern slicks, such as New York's *Vanity Fair* and Baltimore's *The Smart Set*, which ran on advertising revenues, one could get noticed *and* well paid.

Living was cheap and New York was the center of the publishing industry where young writers could meet agents, publishers, and other authors, even get paid for reviewing books.[18] But writers had to make some kind of living before they could live on the proceeds of their writing.

Off-Hours

During their off-hours, when they were not writing or working, writers congregated in the social centers of Greenwich Village—two saloons called the Hell Hole and the Working Girls' Home. The first was "tough and dirty." In the basement lived a pig fed on scraps.[19] The habitués of the back room were small-time gamblers and petty thieves, but visitors included actors and writers. Cowley recalled a gangster who took pity on Eugene O'Neill because of his shabby clothes and said, "You go to any department store, Gene, and pick yourself an overcoat and tell me what size it is and I steal [*sic*] it for you."[20]

Access to the Working Girls' Home was through a side door; the front door was locked. Here mingled people who had lived in the village before the war and those who had just arrived from college or France. The short-haired women wore the uniform of the flapper illustrated in the magazines, and the long-haired men were tweedy and unpressed. There were the uniforms of the individualists and the inconspicuous American middle class. There mingled the cynical, broken, political, moral, artistic, and religious rebels whose ideals had disappeared into the bickering of the postwar conferences on the future of the world, the steel strike, the Palmer raids, and the

Centralia massacre of November 11, 1919, when workers of the IWW clashed with the reactionary American Legion, leaving six dead. The denizens of the Village knew that fools and cheats ruled the world; everyone was selfish and could be bought.[21]

The younger Villagers simply stayed home, ignored their elders, and

. . . . were content to sit in the kitchen, two or three young men with our feet on the bare table, discussing the problem of abstract beauty while we rolled Bull Durham [cigarette tobacco sold in cloth pouches] into cigarettes and let the flakes sift down into our laps.[22]

Ruth

Ruth Lechlitner graduated from the University of Michigan in 1923 and taught high school English in Blissfield, Michigan. A summer vacation in the Rocky Mountains and the Southwest inspired her to apply for a teaching position in a Farmington, New Mexico, high school. In 1924 it was "a small western town full of cowboys and Navajo Indians . . . dominated by seven churches full of godly school board members," one of whom got her fired for lending Flaubert's *Madame Bovary* to a student. But Ruth didn't care; she was on her way to the University of Iowa.[23]

John T. Frederick and Frank Luther Mott had read Ruth Lechlitner's work in *Poetry* and other national magazines as well as their own *Midland* and invited her to join them in May of 1925 to assist in editing their regionalist little magazine.[24] She could also work on an MA in English.[25]

When Paul Corey met Ruth Lechlitner during that visit to Iowa City to complete his novel, he was impressed by her and her poetry, but he was resolutely on the way to New York.[26] Perhaps Ruth would catch up later.

A Misfit in the City

When Corey got off the train, he observed that the streets of Greenwich Village were wonderfully clean and neat compared with Chicago's. He got a job with an insurance company but quit after it became obvious that they expected him to falsify reports.[27]

When Viking finally rejected his first novel, he took a job with the phone company to sign up new subscribers and handle complaints from old ones. He came to the conclusion that "a farm boy with a sound moral upbringing could be a misfit."[28] Being fired from the phone company inspired a short story that Jack Conroy published in his little magazine, *The Anvil*.[29] Corey's third job was with Encyclopædia Britannica, Inc.[30] Paul began working on another novel in his time off.

When Paul's mother died back in Atlantic, Iowa, he was intent to stay out of the ensuing family squabbles. He wrote to his aunt that he was interested in one thing only: writing his novels and getting them published. In a letter dated September 21, 1926, he wrote to her that "my affairs of the heart are progressing." Corey described Ruth to his Victorian aunt as "a mighty decent sort [who] smokes, drinks and all that sort of thing" and was "remarkably brilliant and somewhat recognized over the country as a poet. I think she is quite falling in love with me."[31]

From the time they met, Paul Corey and Ruth Lechlitner corresponded whenever they were not together. On April 4, 1926, Ruth addressed a letter to Paul as "Jurgen," in reference to the hero of the popular novel of that title, a 1919 satire by American author James Branch Cabell.

The hero, Jurgen, is a middle-aged pawnbroker who regains his youth and travels through a storybook land of beautiful women draped in diaphanous garments, comatose princesses awaiting only his life-giving kiss, knights, giants, and kings with magic swords. Lurking behind the scenes is the grand philologist, probably a reference to anthropologist Sir James Frazier, who at the turn of the twentieth century published a comprehensive study of mythology, *The Golden Bough*. In the novel the grand philologist can change a god into a metaphor by discovering that she is merely a solar myth. Politically and in terms of religion, perhaps the most offensive part of the book was the portrayal of hell as a democracy whose demons have voted Satan absolute power for the duration of an eternal war with heaven that must be waged because heaven, in contrast to hell, is an autocracy. The book was the subject of a two-year obscenity case that served only to promote its popularity. It is clearly the model for most of the rhetoric of these early letters, including Ruth's seemingly self-depreciatory remarks about her gender.

117 E. Market
Iowa City, IA

Easter Sunday
[Apr. 4, 1926]

My dear Jurgen:

I am most glad that you, in matter-of-fact no. 13 Bank St., New York City, should be opening magic casements, a little forlornly, upon who knows what perilous seas. It is very charming of you, and altogether in character.

I am grateful for your passing romantic mood; nevertheless, being feminine, and perverse, the least bit piqued that you chose to let me know you have so many Ruths in Iowa to remember. But I forgive you, since moonlight may account for much; and one who performs an autopsy upon the elusive fancies bred of cigarette smoke deserves to be forgiven many things.

You do me a little wrong in thinking that I forget too easily: it is sometimes best, you know, to approach softly with a prayer upon the lips, the templed aestheticism of memory.—Or, if the temple door be guarded, a little star to look at might do one just as well . . .[32] But you, strange Jurgen who dwells within earshot of the call of the muezzin, doubtless prefer prayer.

I envy you South Ferry—I envy you anything that gives rise in you the immortal desire to write beautifully of beautiful happenings. These streets of Iowa [City], Paul, are today murky with an Easter patchwork of yesterday's snow: even the dusk is not too kind to them. So I must work tonight, and forget if I can the promise of a beauty that may come too soon—and the unhappiness that I swear perennially shall not be but somehow always is . . .

Such meanderings are neither discreet nor ethical; only a romanticist could be expected to take them for what they are worth. Perhaps we who live under the shadow of illusion have need for latticed windows . . .

And now to E. A. Robinson, who at least found some solace, however barren, in a gospel of futility.[33]

Ruth[34]

The reference to the guarded door of the temple of memory may refer to a pre-Iowa or at least pre-Corey love affair that Ruth commemorated in a poem published in 1927, "Wife-Thought."

It is the same now, it is the same.
I can be young again; this I can be,—
Breath over loveliness, wind over sea.
Summer remembers, and I shall know how
Leaves bend to purple under the bough.

I can forget him; I can forget
This hollow-crushed pillow once warm to his head.
I am a girl lying still in my bed,
Counting the poplars all silver of limb:
Knives against heaven, shining and slim.

No child cries out for me; I am alone.
My lips to silence sweet in the night;
The dreams of a girl are mist over light
One star to my lover . . . bright stars, shine for me!
(God, he has gone to her: how can it be?)[35]

Ruth received her MA from the University of Iowa in 1926 and went to stay with her parents in Lansing, Michigan, where they had moved after she finished high school in Mishawaka, Indiana. It was from her parents' house in Lansing that she wrote to Paul on August 1, 1926, that she "would like to be able to write poetry that is swift and clear and honest, instead of this bewildered fog of words."

> As for people—I think sometimes that the world finds it necessary to renew itself by propagation; and may the gods pity us for talking glibly about starlight and eternity, when our real business seems to be the eating of potatoes—mashed or fried—whereby we may raise the next generation. . . .
>
> Well, I won't say any more. Chiefly because it is far from becoming and proper for a woman to try to philosophize; we women should know our limitations, and content ourselves with an endeavor to be merely charming![36]

On August 26, 1926, she wrote to arrange for Paul to meet her at the train station in New York and said, "To know that you want me to come will make my trip there mean much more to me." They lived together and shared housework and expenses, but Paul retained a room "just in the event her family should put in a surprise visit," since by October 1926 their relationship had become intimate, apparently unbeknownst to her family.[37]

Ruth found work with several publishing companies, including Harcourt, Brace, Henry Holt, and Farrar Rinehart, before she got a job as assistant to Freda Kirchwey, managing editor of *The Nation*, for which she did some editing, some writing, and read manuscripts.[38] At the same time, she was reviewing books and writing poetry, some of which she sold.[39]

The Uprooted

It was quite in keeping with the times that Ruth should join Paul in the revolt against the village that was spreading across America and that they would go to New York to partake in modernity.

> . . . the homeland of the uprooted, where everyone you met came from another town and tried to forget it; where nobody seemed to have parents, or a past more distant than last night's swell party, or a future beyond the swell party this evening and the disillusioned book he would write tomorrow.[40]

The First World War was over. Some people in the Village were returned veterans of the foreign ambulance corps in France, Italy, or elsewhere; some had been in the armed forces of the United States or its allies.[41] But Paul wasn't among them. He was still in school during the war, but he was surrounded by such figures in Greenwich Village when he arrived. If he did not share their "monumental indifference," he did share their spectatorial attitude.[42] The letter Paul wrote to Ruth when her father died in 1930 illustrates his notion of the relationship between event and memory, the substance of his writing:

> I was much surprised when I read your wire. I didn't suppose the end was so close. After the nightmare of these next few days has past [*sic*] a memory will be born that will be far more genuine and real than the tortured living thing. That is the thing that counts. There is no time for lamentation. At this period the living must nourish the birth of memory.[43]

The First Abortion and the Other Woman

Ruth became pregnant in 1927, but Paul didn't offer marriage and she didn't want to hear that he'd married her because she was pregnant. With a good-paying job, she could afford to find a good doctor, and early that year she had an abortion in his office with anesthetic and postoperative care, rare luxuries in the days before abortion was legal.[44] From this Paul concluded that she was a career woman who would never have a child to interfere with her creative work.[45]

In late May 1927 Ruth returned to her parent's house in Lansing for a visit of about a week.[46] There her college friend Dorothy Hall wrote to her on May 26, 1927, from Pittston, Pennsylvania:

I'm sorry you are unhappy, Ruth, but if it will bring you here to visit me I can't be so sorry as I otherwise would. I'll be delighted to have you—a breath of New York to a homesick soul. Why not plan to spend the entire week-end with me—come Thursday and stay until you have to go back. Let me know what road and what train you will come on and I'll have the brass ford out to meet you—meaning the flivver.[47] I can hardly wait until you get here.

And I can sympathize if I can't do anything else. I've been unhappy, too. Sometimes when I knew why—but oftener and more unhappy when I didn't know. Then it comes with terrible pain, but so vaguely I am helpless. What is it, Ruth? I've tried and tried again to answer myself, but I never can. A restless yearning that has no name, no reason, no answer. I am like you in feeling the need of other companionship beside that of men. I have to have it. Sometimes I pay quite a price to get it. Living with Nadya was such a price in many ways. But I have lived with girls much worse than Nadya, so that it was not as big a price as it seemed to other people. However, I do not intend to ever live with her again. I like her better when she lives far enough away to be less familiar. But no man can take a woman's place in friendship—not even as effeminate a man as Paul Allen. After all, he is not a woman. He does not have the same views, affections, peculiarities, etc. I'm not sorry, of course, but as much as I care about him, he isn't enough.

But it is useless to say everything when I am going to see you so soon. Please let me know just when I can expect you. I haven't heard from Paul Corey in ages. Did you get the letter I wrote you last week in which I went into detail on the subject? Of course I won't say anything to him about things you don't want him to know. I think that I'll write him a note tonight, tho, jacking him up in his correspondence.

Lovingly,

Dottie[48]

Toward the end of May, Ruth wrote to Paul that she couldn't sleep, was having terrible dreams, and woke with such a heartache and sense of apprehension that she felt it would be better not to sleep. She continued:

> But I mustn't write like this to you. Please forgive me once again. You're <u>such</u> a good scout, such a friend, so unbelievably fine. I say over and over what you taught me—"Divine love has met and always will meet every human need"[49] and pin almost every hope I have on that. I don't know what's at work within me, I don't know what will come out of it; but whatever does, Paul dear, know that it must be, and stay my friend . . .
>
> Paul, I've never been other than as straight-forward and as is humanly possible with you. You know I've been your friend, and intend to be your dearest friend so long as you'll have me. Nor do I have any intention of playing the "wanton slut" with you. You can forget that part of it, and it will be better for you if you do. Nor do I have any intention of going to the Devil, by fast or slow route. . . .[50]

She then discussed her return trip with the visit to Dorothy along the way in Pittston. Paul answered, "As you tear down your defense-wall of fear I shall take the brick from your hands and build a shrine of trust. And God pity the being that despoils the tapestry or image of that shrine. All that should be forgotten will never be remembered. It is done! We shall be happy!"[51]

Ruth next wrote from Dorothy's place in Pittston, "Dot and I are planning to go to bed early (yea, Lesbia!)"[52]

Such relationships between women were quite usual during the 1920s and 1930s,[53] and this one was fondly remembered by Dorothy, who fifty years later wrote tenderly of photographs from 1923, added a catalog of her children and grandchildren, and in closing wrote only, "September Song," in reference to a 1938 love song celebrating young love.[54]

Earlier she had written:

> I was such a hedonist that I had no motivation but to enjoy myself and share that enjoyment with anyone I met along the way. I had no social conscience, nor any conservative frugality that would impel me to husband my limited resources. It was not so much the prodigality of a Dorothy Parker or an Edna St. Vincent Millay as it was of the character in a poem of older vintage—Evangeline—
>
> "So was her love diffused, that, like to some odorous spices,
>   Suffered no waste nor loss, though filling the air with aroma."

But it took me over five lively decades to run down. . . . or burn out . . . or whatever it is that happens to beat-up old Dionysians.[55]

On June 27, Ruth was back at work at *The Nation*, and Paul had traveled upstate with some male friends. Ruth wrote to him:

Dotty Fontaine called me up this morning and wants me to come over to her house Wednesday instead of tomorrow, as she is having another girl over then and wants me to meet her. Tomorrow night then, mi [*sic*] love, I work. Tonight I dedicate to the cause of Lesbianism and chiefly long slumber, the latter which mine spirit mostly craveth. . . .

What to feed the addict of Lesbia tonight, I do not know.[56]

The Damned Thing Won't Write Itself

Two days later, Paul wrote, "Vacations always get on my nerves and this one is succeeding remarkably."[57] He continued, "My novel seems to make little headway. The damned thing won't write itself and I feel so tired out most of the time that I never get to it." Paul began collecting rejection letters for his second novel, *Tears of Things*, in 1927 and 1928 as it made the rounds of Simon and Schuster, J. H. Sears, and Coward-McCann Publishers.[58]

Young writers learned that in France they could live even more cheaply and, with some savings, wouldn't have to fit their writing into deadening work schedules. They would be free. Unlike the returned veterans of the First World War, the younger intellectuals, like Corey and Lechlitner, "went streaming up the longest gangplank in the world; they were preparing a great migration eastward into new prairies of the mind."[59]

Late in July of 1928 Ruth visited with her family in Lansing and wrote lovingly to Paul. On July 22, Paul wrote of walking to Sixth Avenue on an errand with a friend and investigating a crowd gathered by the Jefferson Jail:

A man had fallen down in the street with a heart attack. Someone had carried him to the walk and propped him up against the jail. He twisted, squirmed, writhed—clutched at his heart with both hands, choked, panted, gasped. He chewed his fingers till the blood came. A cop was waiting for the ambulance, looking sad.

The two continued on their errand and returned the same way to see that the ambulance had not arrived.

It was plain curiosity. I thought he might die—I have never seen a person die. The bag of groceries was by the fellows [*sic*] elbow. He looked like a dog that has been hit by a taxi and people watched him as if he were a dog. The ambulance finally came.[60]

Obviously, he had not yet told Ruth about the death of his oldest brother in an epileptic seizure.[61] In his letter of July 25, Paul asks Ruth to give his regards to her family, so they must have announced their marriage to her family.

Also I am sending you an affidavit form which you should have filled out to certify your birth. A bit of good news, you can go on my passport which means only one passport charge and one visa fee. Cut down cost twenty dollars on the initial payment at least. We will have to bond the typewriter on entrance to France but the money will be returned when we leave. . . .

Edith arrived—she bought a quart of gin or had Karl buy it and we got plastered. Had to do something. She was present. Last night Mantz invited Templeton, his wife and myself down to tea (Tea Mantz style) We talked there until eleven thirty before we were aware of it. Harry got gloriously plastered . . . two nights in succession is a little too much for me.

I was up at Nelson's Thursday evening to hear the fight. . . . Well, five more weeks.

. . . I haven't been able to type any this whole week. God damn, I'll be glad when I can get away from people.[62]

Charles Nelson was Paul's high school friend and college roommate who later would buy the land they would live on in Putnam County, New York.

On July 26, 1928, Ruth responds that it is "good news about the passports and thanks for sending affidavit blank [to affirm her birth]." She says she thinks the episode of the man dying of a heart attack would be a good nucleus for a short story that magazines would buy and mentions *New Masses*. "Why don't you try it, dear? And let me see it when I get back. Beginning tomorrow, I go on schedule for 2 hrs. of work writing every day. That is, if the darn family lets me alone that long."

Paul answered that he was pleased at what she said about the dying man incident. "It gave me one of those momentary flashes of self-assurance—a drop of oil, a very tiny drop, on a dying flame. It wasn't more than a flash, however, for I remembered things you've said on several different occasions . . . Anyway, I appreciate your wanting to make me feel good even if

you couldn't mean it. . . . I'm glad you've been getting some writing done."
He continued:

> Darlingest, I'm in an awfully bad way. Things are getting worse and worse.
> I'll be so glad when you get back. Sometimes I find myself wondering who
> you are, or just why it is there seems to be something gone. Again, you
> don't exist at all and the only thing I am conscious of is a great emptyness
> [*sic*]. I'm so sick of everything. Last night I lay awake, oh for hours. My head
> went around and around, my face burned like fire. I almost called you on the
> telephone—I almost went out and jumped in the river.[63]

In the same letter he describes his work with Encyclopædia Britannica:
"It's like a vampire, Ruth, a wierwolf [*sic*], it's sucking all the life and vitality
out of me. God Damn! that filthy hole of decaying old world culture stirred
by those slimy fingers of new world idiots." He adds:

> You must remember to get your teeth fixed and a new extra pair of glasses.
> Go easy on your money, sweet, then maybe I can get out of that hell hole
> sooner than I expected and still we'll have what we planned. I shouldn't talk
> about getting out before you do, but I can't stand it.[64]

Ruth received a letter from Dorothy Hall dated July 10, 1928, announcing her
marriage:

> The role of perennial <u>fleur du mal</u>[65] is enervating, so I prefer to abandon it
> in favor of comparative domesticity and let F[rancis Quick] do the worrying
> over the consequences of my wayward antics. . . .
>   When are you and Paul leaving for Europe? Give him my love and tell him
> that in the end we all acquiesce to socio-biological regulations and that one
> of the most recalcitrant rebels had gone over to the enemy.[66]

Bohemia, Cowley wrote, could exist only in a capitalist society because
it is a revolt against industrial capitalism.[67] It was that same industrial capitalism against which they revolted whose acceptance and approval they so
eagerly sought. The younger intellectuals, like Corey and Lechlitner, were
joining the exodus to France.[68]

> The village was almost deserted, except the pounding feet of young men
> from Davenport and Pocatello who had come to make a name for them-
> selves. . . . who came because there was nowhere else to go.[69]

Perhaps Cowley had not heard of Marne, Iowa, Mishawaka, Indiana, Iowa City, or Lansing.

France

In late August 1928 Paul and Ruth went by ship to Le Havre, then to Paris, and biked south from there.[70] They hoped to "find French peasant life in all its naturalness."[71] Appreciating the "old-world stillness"[72] and "bits of half-forgotten of history,"[73] they explored churches, cathedrals, chateaus, renaissance hunting lodges, palaces, and villages. They lunched on French bread, cheese, and wine. They visited markets, listened to remembrances of the recent war, and gazed at colorful landscapes and sunsets. They ducked under trees and awnings to escape rain showers and sometimes gave up and cycled through the rain. They rode along bumpy muddy roads and found a level canal tow path where they could relax until they had to dodge a barge-towing donkey.[74] They watched peasants plowing behind their plodding, stoic white oxen and antiquated plows. They labored up and flew down steep mountain slopes. They mused on the ancient and the modern as they looked at an idling tractor beside a team of oxen and wondered where the people were.[75] They gratefully collapsed into the beds of local hostelries along the way.

They rode past new industrial complexes and rocky villages until they all began to look the same. They cycled against headwinds, sped with tail winds through terraced vineyards and broken stone fences; through olive and orange groves. Ruth's hands were numb and her fingers so cramped she couldn't hold down the hand brake.[76] Finally, they topped a ridge and saw the sea, "as lovely blue on this day as tradition has ever painted it."[77] At last they reached Bandol-sur-Mer and celebrated with a bottle of Provencal white wine in a café as they watched the fishing boats in the bay.[78]

They found a real estate office where someone spoke English, and they paid half a year's rent for a place across the road from the sea.[79] They were on the French coast and free to write, but Paul was despondent about his prospects.

> During the months we lived in Bandol, I got little writing done. There was no reason to be optimistic about my becoming a writer. I began to feel that perhaps I was like others I had known who said they wanted to write, but after they got married and had sex regularly, they never got around to writ-

ing. They took a job, started a family, and the dream of becoming a writer remained only a dream.[80]

Or, as Malcolm Cowley put it, "Here in this ultimate refuge there were no distractions whatever, nothing to keep them from working except the terrifying discovery that they had nothing new to say. Boredom and loneliness set in."[81]

Fun on the Farm

They had been in France less than a year when a telegram from Paul's University of Iowa roommate, Charles Nelson, arrived to initiate the third phase of the writer's life that Cowley described:

> Bought farm on condition title is guaranteed within month. Letters follow. Cheerio.[82]

A few days later, the letters began to arrive:

> How'd you like to come back this summer? I think we could have a lot of fun. . . . We're going to dig some place for garden Sunday, day after tomorrow, and maybe begin to clean out the spring.[83]

And a couple of weeks later:

> I shouldn't particularly rush home if I were you, for the farm will "keep" and you may not get back to England for some time. On the other hand, I know I'd be anxious to see the place if I were in your place, and I do think we could have a good deal of fun working up there the coming months if you returned.[84]

Paul Corey would later write two books about what happened next. One was a manual, *Buy an Acre*, complete with detailed plans, instructions, and budgets for how to purchase a small plot of land, build a chicken coop and house on it, grow food, and be somewhat independent of the surrounding urban economy. The second would be a fictional version where we learn Corey's idea of fun, the sense Charles Nelson was referring to in his letters. The fictional family of the novel, *Five Acre Hill*, includes a father, mother, son, and daughter.

> When the meal was well along, Bob [the father] looked across the table, eyes twinkling, and asked: "Moms, what did you do today that was fun?"
> The others stopped eating, their mouths forming circles of astonishment. All of them knew that "fun" had a special meaning to Bob. A game

was a game to him—energy expended. You worked off your surplus energy and toned up your body playing tennis or swimming, but that was all. You agitated the air or the water with your arms, but when you quit, the air and water returned to normal and there was nothing to show for your expended energy.

Fun was different. You expended energy, you toned up your body and you built something, you created something. You were changed and the world round you was changed, and you enjoyed what you did from the top of your head to the tips of your toes. That was having fun.[85]

Granville Hicks wrote of Hemingway, one of the older generation of writers who had experienced the First World War, that the remedy for the disease of his generation was action:

> . . . if all else fails, if social obligations lose their force, if the desire for success is dead, if all philosophies seem equally meaningless and all philosophers are equally futile, action remains, not action for a cause but action for its own sake, the unthinking, unhesitating, and if possible hazardous exercise of the body.[86]

Such pointless action Corey saw as an evasion of responsibility. For Corey, to be meaningful, action had to have a consequential and tangible result that one could appreciate locally, such as building a house. Not for him were bullfights, boxing, deep sea fishing, or big game hunting, but the physical activity of construction, building, growing food, tending chickens, providing for his family.

But it could be hazardous nonetheless. The editor of the *New Masses*, Isidor Schneider, wrote to Ruth regarding some revisions and opened with "Sorry to hear about Paul's accident, and I hope by the time this reaches you, any worry you have about him will be over."[87] A friend wrote to Ruth in 1936:

> Dog-gone it—Paul ought to leave axes alone. I don't believe he's got the right background to fool around with things of that sort. I hope he's better by now and the book is in the hands of the publisher.[88]

Return

Paul and Ruth returned to the ground Charles Nelson had purchased just across the river from West Point at Cold Spring on Hudson. They paid for their part of the land with a small inheritance from Paul's mother and the money they had saved.[89] Paul knew from his experience that he could not in the long term both write and hold a full-time job. Their plan was to build a dwelling and grow most of their own food to be as independent as possible from the larger economy. That way they could devote part of the day to work and part to writing without the draining burden of working for an encyclopedia or a real estate magazine.

Many of the other exiles returned from Europe to New York City to find places somewhere in the greater writing industry of advertising, journalism, editing, and publishing. They longed for their lost freedom and dreamed of retreats in the abandoned farms of the Connecticut countryside. The land was available because the farmers had been mining the soil for generations with no eye to regenerating it, leaving only the stones and thistles as markers of where their farms had been. The freedom-seeking city folks bought the played-out farms and opened at least summer houses where they held parties and commuted to work in the city. Thistles and brambles took over the abandoned fields because the new residents had no relation to the land.[90]

In this Paul and Ruth were determinately different. They not only built the structures they used but also cleared the land and grew the food they ate. Meanwhile, both continued to write. Ruth kept her job in the city, and Paul wrote short stories and began to think of his new novel, one of greater scope than he had before imagined.[91] In this he was not alone because, as Cowley explains,

> What the exiles wanted to portray was the lives and hearts of individual Americans. They thought that if they could once learn to do this task superlatively well, their work would suggest the larger picture without their making a pretentious effort to present the whole of it. They wanted their writing to be true—that was a word they used over and over—and they wanted its effect to be measured in depth, not in square miles of surface. . . . They wanted to build smaller structures, each completely new but with the native quality of New England meeting houses or Pennsylvania barns, each put together with patient pride, each perfectly adapted to the life it sheltered.[92]

To do this, they would have to return to the kind of regionalism they had revolted against when they first came to New York.

4

LIVING ON THE LAND

Realism in America grew out of the bewilderment, and thrived on the simple grimness, of a generation suddenly brought face to face with the pervasive materialism of industrial capitalism.

Alfred Kazin, On Native Grounds, *15*

Back to the Land

Paul and Ruth moved onto their land just before the crash of 1929 became the Great Depression.[1] They took the train along the Hudson River to the Cold Spring station. Their farm had long been abandoned and taken over by weeds and birch saplings, though the broken-down stone walls marking the fields were still visible. Charles Nelson and his wife, Kay, had moved into the main house with the belongings Paul and Ruth had left behind when they went to France. The shell of another cottage was empty. After a few days, Paul and Ruth traveled to the Midwest to visit their families and collect possessions they had stored. When they returned to Cold Spring, they spent the rest of the summer making the cottage habitable.

But they needed money for building materials and for other expenses they would incur in their move back to the land. In September they moved into a one-room apartment in the city and found writing jobs; Ruth with the *Book Review Digest* and Paul with the *Real Estate Record and Builders Guide*. While Ruth had long practice in writing book reviews, Paul was not enthusiastic

DOI: 10.5876/9781646422081.c004

about writing stories about real estate sales. Paul's attention was focused on his plans for building a house to make their rural property a base for writing so he would not have to work at such a job.

By March 1930 they had enough money to buy a Model A Ford pickup, and by the next month they were ready for the move. Paul's coworkers were astounded that anyone would give up a job in the midst of the increasing unemployment. To play into increasing fears of communism, Paul told an inquisitive colleague that he was moving to the country to "pour concrete slabs for Russian siege guns." The colleague asked the boss whether he should report Corey to the FBI.[2]

During the summer, Paul built the shell of a new cottage while Ruth commuted to her job. Paul returned to the city in the fall to a job working on an encyclopedia with the Crowell Company for a salary of a little more than half of what he had been paid before.[3] In April he quit that job and returned to Cold Spring. Another Iowa friend was Charlton Laird, a Columbia PhD student. He, the Nelsons, and Paul built a small cabin for the Nelsons, whose old house on the property had burned down. They would live there while building a larger house, and the cabin would be a garage for the new house. When Paul and Ruth moved into the cottage he'd built, the cabin they'd been living in was vacant, so they rented it to Charlton Laird and his wife, Floy.[4] All of them remained lifelong friends.

With the cabin completed, Ruth negotiated to come into the office only once every two weeks so she could live on the farm and write there. The writers would write in the mornings and build in the afternoons. Their work talk became the basis for a satiric story that Charlton, Paul, Ruth, and Floy wrote on the violence of gangsters and speakeasies, "Machine Gun Culture." *Vanity Fair* published the piece under the title "Bullet-Proof Culture" in 1931 and paid $100,[5] about $1,613 in 2019 dollars, Paul's first freelance income. Paul did not find encouragement in his share.[6] The piece was a parody of the burgeoning gangsterism fostered by the Volstead Act's prohibition of alcoholic beverages. What gangster bosses lacked, the piece suggested, was university degrees that they could get at the Volstead Memorial University after a course of study leading to a bachelor of science in racketeering, which would make them immune from being elected to political offices because Americans did not elect college-educated people to public offices.

Industrialization May Have Gone Too Far

In addition to writers, many others shared the vision of fleeing the cities to live on the land. In his 1934 report to the president, Secretary of Agriculture Henry A. Wallace wrote, "Industrialization and the rush to the cities may have gone too far. We may be entering a prolonged period of urban unemployment." He discussed a "back to the land" program, noting that such a movement had been going on for a number of years in New England.[7] Paul Corey wrote that he hadn't started a trend but exemplified one.[8]

While the New Deal sponsored a Subsistence Homestead program, its main thrust was toward industrialization of American farming through programs that rewarded the largest and most market-oriented farmers.[9] The homestead program was short-lived.

Regarding the back-to-the-land movement, agricultural historian Robert West Howard wrote in 1945:

> The home, fixed on a two- or five- or ten-acre plot and owned in fee simple by the uncommon man and his uncommon family, would again become the bulwark of American life. Human roots would sink deep and true into the good earth. The valued inches of topsoil—the one constant, seasonally recurring asset of the nation—would become a common ward instead of a mining property exploited by one fourth of the population because of the cash economy of the three fourths living in cities.[10]

Such a movement back to the land would end "the historic threats of concentration of power or wealth by any one group, or person."[11] The agrarian movement in the United States had two dimensions. The country life movement wanted to make rural people more efficient and sophisticated, to organize them to mechanize and commercialize their farming—to industrialize farming and rural America and make rural people more urban.[12]

Howard exemplified the other branch, which advocated a return to the land and praised the virtues of a more natural farm life. The chief proponent of the back-to-the-land movement was Ralph Borsodi, who left New York City in 1922 to live on a small farm on Long Island.[13] The crash of 1929 started a reversal in farm literature—from escape *from* the farm, as in Corey's first two volumes, to a flight *to* the farm, as in his third. There was a spate of autobiographical and instructional books to this end, including Corey's *Buy an Acre*.[14]

Malcolm Cowley reviewed Borsodi's book, *Flight from the City* (1933), under the caustic title "Homesteads, Inc." and claimed that Borsodi "belongs to the class of those who might be called push-button prophets," who want to remedy the injustices of the world without disturbing anyone.[15] The quality of life in rural areas that Borsodi advocated depended on the plentiful and cheap manufactured goods of the industrial cities. There could be no prosperous rural life without participation in the money-economy of capitalism, as Paul and Ruth were finding out.

Thus, the solution to urban or industrial woes could not be for everyone to move to the country. To address those troubles required a reorganization of industry and the economic system itself. To hold up rural life as a solution was to evade the problems of capitalism and industrialization, not solve them. Unless they sank to the level of medieval surfs, Cowley concluded, homesteaders could never be really self-sufficient because they would need industrial products such as canning jars and lids, electric motors and appliances, tractors and other machines, fuel and electricity, as well as money to buy them and to pay taxes.[16] Such a move could not alter the course of capitalism. So ironically, he pointed out, the back-to-the-land movement depended on the industrialization it was ostensibly fleeing. So, paradoxically, to achieve the rural life free of the woes of cities and industry depended on the products of industry. While this could be an individual solution, it could never be a society-wide one.

At Cold Spring, Paul, Ruth, Charles Nelson, and his wife were surrounded by like-minded people. Most had built their own houses. The two qualities they had in common, wrote William Seabrook in *Reader's Digest* in 1939, were the smallness of rooms and the greatness of spirit.[17]

What they likely also had in common was a low standard of workmanship and unconventional floor plans. In 2011 a writer named Gina Van Norsdall and her family lived in the house Corey built on Corey Lane. She described the house as quirky, with few closets, a tiny kitchen, and small bathrooms. Staircases take interesting twists at the bottom, and there is not a square joint in the house, she said. The contractors she and her husband hired to work on the house commented that they could not imagine how the house had been built.[18]

But to this crowd, work and independence were more important than cash or credit. One example Seabrook described was the young novelist Paul Corey and his wife, the poet Ruth Lechlitner. Corey told Seabrook that the principal characteristics their lifestyle required were disposition to work hard, ability to get along with neighbors and willingness to give up some comforts.[19]

"Willie Seabrook heard how we were beating the Depression without going on Relief," Corey wrote. "He interviewed us, had dinner with us—he loved my wife's cornbeef [sic] hash—and wrote us up for the Reader's Digest [sic]."[20] After a globetrotting writing career, Willie Seabrook settled down in the Hudson Valley close enough to Paul and Ruth to help them roll the bathtub into the new bathroom and celebrate the publication of Paul's first novel along with the house warming for the new house.[21]

Medieval Serfs

For nine years Paul and Ruth devolved to something like the standard of living of Cowley's medieval serfs with no running water or electricity and only a pot-bellied wood stove for heat. She washed their clothes by hand with yellow soap on a washboard in an old bathtub they found in a dump. She ironed with a heavy sadiron they heated on the stove. Paul carried water from a well about two city blocks distant from their dwelling and cut firewood. For light they had oil lamps.[22] Admittedly medieval serfs had neither pot-bellied stoves nor bathtubs for washing clothes. Until they got a kerosene stove, Ruth cooked over a fire in a ring of stones protected by two umbrellas—one for the fire and one for the cook.[23]

As the Depression deepened, Paul and Ruth were more or less free from the constraints of the collapsing economy, though Ruth did get freelance work in New York. Until 1948 she reviewed books for Irita Van Doren at the *New York Herald Tribune* and for *Poetry, The New Republic,* and *New Directions.*[24] In the meantime they ate the vegetables and fruit Ruth canned and preserved during the summer, their garden produce, and their chickens and eggs. They came close to living off the land.[25]

Corey continued to write in the mornings and spend afternoons building a larger two-story stone house or producing food from the gardens he and Ruth terraced after the pattern they'd seen in France. For building material Paul used rocks from nearby stone fences.[26] Ruth commuted by train to her job with *The Nation* in New York City, sometimes spending as much as a month in town, and wrote poetry. The 1930 Model A Ford pickup served for the six-mile drive to town to the train station and to haul building materials.[27]

While Ruth was in Michigan when her father died in 1930, Paul wrote to her about the progress of work on the first dwelling:

> I am trying frantically to get the Nelson Menage moved up the hill by Thursday. They are slowly driving me insane. I've lived in filthy places and have eaten impossible food but never has anything equaled my present situation. The chickens have been moved up to the old tool shed and that relieves the place of some confusion.
>
> Charlton was a good help and we managed to get the house to a finishing state, but the windows and doors haven't arrived as yet. I completed the roof this afternoon and got half the cement surface on the floor. So far the Nelsons have been paying for my food which I think is quite fitting and proper. I've got to suggest paying come again [*sic*] soon but I hope to do it in a way that will make them refuse to accept. What I want to do is get the place all fixed up and them out of here by the time you get back.[28]

View from the Hudson

One person's paradise of the gospel of work is another's canker sore on the landscape of the gospel of wealth. Not everyone was pleased with people moving to their scenic Hudson Valley to homestead. In his manual *Buy an Acre*, Corey devoted a chapter to "The Community." When you first come to your land, he cautioned, you are a stranger from the city, perhaps tainted with liberalism. Perhaps you belong to a union or believe in equal rights for black people.[29]

The local politician will rightly be worried because his family has married into the community for many years, creating the ties that guarantee him the votes he needs, and his relatives have connections. And those connections have others. "That's how he gets elected."[30] He can't depend on you for your vote. And there are several hundred suspect folks like you moving into his area. Maybe you thought your tax assessment was high and invited the assessors to visit your place to see for themselves. Tell them, "I just figured that you gentlemen ought to see my land before you assess it. I just want to make sure I'm assessed as much as the next fellow, that's all . . . But no more."[31] The assessors will never have encountered anyone who wanted to pay fair taxes. To win the merchants, pay cash.

The president of the flower club might be horrified that you're living in a chicken coop while you build your house. "Who was responsible for this terrible shack cluttering up her beautiful countryside?"[32] Her gardener raised the flowers that won prizes for her, and to her your little shack is a monstrous blot on the landscape. She starts a campaign to keep you and your kind out of the area, but you won't hear about her crusade.[33] It will be private. Corey mounted his own public letter-writing campaigns to newspapers, businesses, and politicians.[34]

One of your neighbor's cows gets into your garden and you soon learn that the natives think their livestock can do no wrong, and you decide to tend your livestock well so they won't trouble neighbors. People see the progress on your place, and the local newspaper does an article on it. The president of a women's political club has her chauffeur drive her to your place to invite your wife to join. She comments on how quaint the house is, but your wife is too busy gardening, building, canning, and doing useful things to join this woman's organization. Madame's chauffeur drives her home where she agitates for a return to the good old days when city people stayed in the city and the people in country estates could enjoy uninterrupted vistas of picturesque landscapes.

You may join other homesteaders to organize a cooperative or push for better roads and schools. Then you're a radical, maybe a Red. But when your kids are in school, riding the bus with the other kids, and you keep paying cash, you'll have a place in the community.[35] In *Five Acre Hill* Corey dramatizes such events.

Last Frontiers

Paul was certainly willing to work and give up comforts. Ruth captured one vision of the phenomenon in her poem "Last Frontiers," published in the communist journal *New Masses*:

> There were lands once to the West virgin and waiting:
> When the near acres were sucked dry and the old men
> Wisely in famine perished, then were the young driven
> By hunger to the far valleys, strong-limbed, to reap, to mate
> Under lark song and prairie wind. And so sufficed
> The earth-taste to men's bellies, so were made abundant

The wombs of their women. (There were dreams in your bones
O pioneers, that we have long forgotten!)

New seed to that soil; but into the heavy furrows
Fell other and stranger seed: sweating blind under the sun
Men poured into the young corn their heart's blood
That we might harvest golden alphabets.
So from our eyes the light from those wide fields perished:
In the shadow of brick walls we bent, book-wise, forgetting
Wind-syllables in the long grass, leaf-words in spring.

We slept apart from the earth; we knew hunger
Not of the flesh, and we set our feet upon the highways
Toward those frontiers of shining glass and stone—
There were we driven, there indeed you have seen us:
Boys in the grey day stale from rented bedrooms,
Boys with pale hands and flaccid loins
Chained to thick ledgers, bound to polished desks;
Girls with hard eyes and calloused fingertips.

Nervous, sterile-scented, having careers . . .
(Only we never knew that the black steel girders
Were set in prairie corn; we were never told
That the dark furrows bore this leaping stone
To the wings of the highest tower.) And so we came,
So we were used, and so again cast out:
There are too many of us in these days of famine.

We are the young uprooted: where are the new lands
Waiting us? (Let us think now of our elder brothers
Who had no time to question before they were answered
With the points of bayonets: shall we too be answered?
Let us think now of our younger brothers
Idle on the home steps, their yet untried hands pleading
And empty; their untried fresh minds wasting. . . .
Whose need rots their young bones: shall they too be answered?)

We remember our books—and somewhere under the sick breath
Of old men, somewhere beneath the patriotic dust

There were new words that leap to our minds like fire
Now we recall them: words that shall serve us now
In place of bread. Regenerate in revolt
We shall rise up, create our own new lands,
For the last frontiers are taken.[36]

The Gospel of Work

Corey emphasized the virtues of hard work, working for one's self, being independent. This was the old American Puritan culture that demanded character,[37] values that Greenwich Village was famously revolting against.[38] This outlook is the gospel of work, that work creates all value, as opposed to the gospel of wealth, that wealth produces wealth. The first was the perspective of the self-sufficient people of the turn of the nineteenth century. The second provided the ideology of corporations.[39]

In 1889 Andrew Carnegie had proclaimed his gospel of wealth—that capital produces wealth by natural processes that the new science of economics would elucidate. It was the responsibility of more highly evolved beings like Carnegie not only to manage wealth but to be sure it was used for good purposes. By endowing university positions in economics and promoting the discipline that justified his operations, he and others like him started a great cultural revolution in the United States.[40] The gospel of wealth and economics provided the ideology that would support the rise of corporate America over the next century. These preoccupations replaced the earlier Puritan ones based on hard work.[41]

This corporate ideology began systematically to displace the gospel of work.[42] The new order was one of managers, experts, scientists, and engineers, all dedicated to bringing efficiency and reason to corporations and government. When Corey was on the staff of Encyclopædia Britannica, one of his coworkers introduced him to Howard Scott, who created the concept and term "technocracy." Over dinner in Greenwich Village, Corey expressed reservations about engineers running society and didn't get a very satisfying idea of where writers would fit in.[43]

"Many who attack the modern world are clearly unhappy living in this [twentieth] century. They have little sympathy for anything that is for the masses, seeing always some sort of fascism or Stalinism around the corner,"[44]

wrote cultural historian Warren Susman. The universal abundance of a successful capitalist system would annul want and suffering and any necessity
for alternatives to capitalism such as socialism.[45] Looking back on the Great
Depression from the vantage of the Great Repression in 1954, it appeared
that the New Deal had solved all of the problems with capitalism, that history had proven Marxist doctrines to be false. "Instead of the rich getting
richer and the poor getting poorer, there has been a remarkable leveling of
income, and Big Business, however reluctantly, has accepted government
regulation."[46] The lessons of the Great Depression had been learned.[47] Both
Hicks and Susman had underestimated the power of the cultural revolution,
but perhaps not the power of the owners of capital to undo the New Deal.

The alternative to participation in the mass society and its ills was to
withdraw from the burgeoning technocracy, the world that belonged to
the managers, engineers, and corporations,[48] into the kind of space Borsodi
advocated, a return to the land and the practice of the gospel of work where
what people produced counted for more than the money they made. For
people of that previous, Puritan ethic, money corrupts whereas work sancti-
fies.[49] This was a theme in Corey's writing and his life.

Ruth Lechlitner wrote of the transition and the distinction in her 1936
poem "Interview," published in *New Masses*:

> *In his museum village I found smiling*
> *The great industrialist, soft-palmed, soft-voiced:*

> *"No man should work for profit only"*

> *Into the past he looks—*
> *The standardizer, modern mass producer—*
> *Admires the craftsman*
> *Carving with slow care*
> *Shaping with loving pride, fine hands*
> *The single object, part by part, the whole.*

> *Eight smoke stacks bar the sun,*
> *Smudge the flat basin of the River Rouge.*
> *Bones bent equally under the loud drums*
> *Of overhead conveyors, hands stained alike;*
> *The sweat of armpits to the sweating steel.*

Eyes, nostrils thick
With the same grease-heavy air. . . .

"In money for its own sake
I have no interest"

From a year's labor of these numbered workers
(Bread at the day's end, soggy sleep)
Profit enough to buy imagination:
One free man's dream, one life
Memoried in oak: carved chest, museum-piece.

On the assembly line
Hard shoulders, stout wrists, quick hands:
Steve Projak, turning
Bolt after bolt on black connecting rods
One way one hundred times
An hour eight hours a day
Six days a week months years
(Wife and five kids in Three Oaks
A raise on the speed-up
Partner laid off in June)
Almost enough to eat, to sleep,
Bolts for connecting rods: steel charity.

"The only
Charity I know is paying people
Fairly for what they do"

Steve Projak, your ancestor
Might take such oak as grew
Once by this oily river:
Shape with mind and heart
Leaves under sun and moonlight, forest shadow,
The rose garland, or the rich-lobed
And rounded flesh of fruit.
You earn your bread, Steve Projak,
Fairly for what you do.

"It is only by the exchange
Of benefits that profit can exist"

We talked of the Plant's efficiency, shook hands.
His sentiment sure antiseptic to
Conscience decayed, his lean face phosphorescent
In the green light of evening, I left smiling
The soft-voiced visionary, great industrialist.[50]

Early in October 1935, Isidor Schneider wrote to tell Ruth she would recommend the poem at the next *New Masses* staff meeting but requested one change:

> One phrase I don't like and ask you to omit—the phrase—"all art is bunk." It seems incongruous to find it in a poem. It is sectarian.
>
> The revolutionary movement certainly would not promote writers' congresses and art congresses on a national and international scale, would not announce itself as the defender of culture, if it thought all art is bunk.[51]

In his life Corey exemplified the gospel of work; in his writing he extolled it.[52] But he also understood the uses of money, and throughout his life he kept meticulous accounts. His two books about homesteading include detailed price lists. He has the father in his juvenile novel *Five Acre Hill*, say:

> "I did a little figuring today," he [Bob] said, sweeping the family with a triumphant glance. "It cost us no more to live out there this past summer than if we'd lived right here in New Delphi [where his job is, twenty-three miles from Five Acre Hill]. I sent off a final check to the Wayne Lumber Company to square our account and we'll have twenty-five dollars of our ten-dollar-a-week accumulation. We don't owe a cent on Five Acre Hill as it stands. We've put seven hundred and thirty-five dollars into it, and I figure the place is now worth two thousand. . . . That place is worth a lot more than we put into it and we don't owe anything on it, which is the important thing."[53]

For perspective, $735 in 1945 was equivalent to $10,594 in 2019; $2,000 would be $28,826 in 2019. Throughout his life Corey practiced both perspectives with an emphasis on work over money. He wrote, "I had a dream of building a mutual relationship upon an economically sound base."[54] He kept detailed accounts to continually check on the soundness of that base. He reiterated these views in detail in his 1950 book, *Homemade Homes.*

Iowa

From the vantage point of his plot of ground in upstate New York, Paul Corey could look back on his life in Iowa where, aside from sporadic episodes of farmer rowdiness that caught the eye of eastern journalists, there were no great events.[55] There was no revolution to watch, as in the newly minted Soviet Union and the struggling China; no hopeless defense of democracy against fascism to move the world, as in Spain; no rise of fascism to frighten the world, as in Germany and Italy. There was just the everyday farm life of mediocrity that Corey had fled. But that was a way of dignity. Hard work was not always rewarded; it had to be its own reward. As well as honesty and truthfulness. A person could see the results of that work, honesty, and truthfulness and stand tall at the end of each day. But that way of life was vanishing.

Paul Corey chronicled that farm life as it was silently bulldozed away and with it the economic cornerstone of American democracy. That, he had lived; the destruction of the lives of hardworking, honest people he had seen and would never forget. And that life he did not want the rest of the world to forget; that life he did not want to see disappear from the pages of history as it would be written by the victors.

Corey saw the demise of family farming not as the failure of numerous individuals because of their bad management or poor market decisions, but as a result of wider social forces.[56] As one of his characters observes:

> It seemed to him that there were three things—himself, healthy and poten-
> tial; circumstances beyond his control which had gone haywire, and in
> between something-or-other called Life. If the things beyond his control
> didn't get straightened out so that somewhere or other his potentialities
> could fit into them; then the thing in between did not exist.[57]

That social understanding of individuals is what sociologist C. Wright Mills, writing in 1959, called the sociological imagination, the ability to see and understand the "the interplay of individuals and society, of biography and history, of self and world."[58] To this anthropology adds the notion of shared patterns of thought that people learn in the process of growing up when and where they do, culture. These patterns guide people's actions and constrain them. Cultures are related to societies, and where there are many social divisions there are as many different overlapping and often contra-dictory patterns of thought. Thus, the difficulty of describing "a culture."

Ethnography allows anthropologists windows into those complexities that also provide much of the material for fiction and drama.

Corey wrote to Jack Conroy:

> For a long time, I've been trying to write stories so that the reader will see groups and get a feeling for groups and an understanding of groups, not just a feeling for John Doe and his belly ache, Mr. Ford and his Millions, Mr. So-and-so and his sex. It all sounds a bit insane and silly and maybe it is.[59]

Later he wrote Conroy, "I made no attempt at a plot story; I was thoroughly conscious in not dealing with a hero, heroine and villain."[60]

Paul Corey's growing up in a mosaic of less than assimilated and very visibly different ethnic groups that were, in Corey's words, "freckles of nationality on the skin of the body politic" might have contributed to his montage method to storytelling.[61] One of Corey's fellow journalism students at the University of Iowa would struggle with the same issue of how to describe objectively groups but developed a very different approach.[62] George Gallup established the method of characterizing groups statistically and representing variability in statistical terms. Both were attempts to give more form to the notion of the "mass," which was emerging with the mechanized industrial period of mass production, public relations, and mass communications that defined an age of plenty and fulfillment for all. Corey's viewpoint, focused on individuals in their social situations, was more aligned with the older Puritan producer-capitalist gospel of work.[63]

While one of these Iowans rebelled against the "standardized modes of thought and behavior in modern mass society,"[64] the other defined it. The Industrial Arts Exposition of 1935, organized to display "industry's present solution of the practical, artistic, and social needs of the average man" and held in the new Rockefeller Center, the embodiment of machine age mass culture, was to be judged by a committee of a hundred average Americans headed by Roy L. Gray of Fort Madison, Iowa, who had been chosen as the Average American.[65]

It was this machine culture, described in anthropologists Robert and Helen Lynd's study of Middletown, that journalist Stuart Chase contrasted with the handicraft culture of the Mexican village of Tepoztlan depicted by anthropologist Robert Redfield.[66] This suggests a third answer to the question of how to describe social orders: ethnography.[67] Anthropologist Oliver

La Farge wrote in 1962, "The greater the numbers, the more complex the society. . . . the more nearly impossible it becomes for the investigator [ethnographer] to view and comprehend it all. The humanistic, face-to-face method comes up against firm limits of numbers, variety, or usually both."[68] The sociologists' answer is like Gallup's, but these data cannot encompass human variety, so there must be interpretation based on observation, ethnography.[69] Corey meant his Mantz trilogy to be ethnographic in the Lynds' sense of the term.[70]

The Second Abortion

Paul and Ruth were carving out writing careers based on the economic foundation of their semi-independent homestead, free from the strictures of routine jobs and distant from the dominant machine culture of the time, more aligned with the handicraft tradition of past times or distant places. With such limited access to funds, the last thing Paul and Ruth could afford was a child. In 1932 the economy continued its collapse, and thirty-one-year-old Ruth became pregnant. Paul said that after all the effort of getting himself into an economic situation where he could write, he didn't want to have to give up his writing and get wage-work to support a child like other couples they knew. Ruth resisted the suggestion, but when he pointed out that their impoverished lives offered little for a child, she relented and went to an abortion mill to terminate the pregnancy. She learned just how different that experience could be on the bottom of the economic heap than it had been when she had plenty of money.

As she sat on a cot awaiting her turn, she heard the women before her incessantly screaming. Then the abortionist looked at her and asked what she was doing there, that she looked like an intelligent and healthy married woman who should be having children, not getting rid of one. She muttered, "I want the baby; it's my husband who doesn't."[71] It was physical and emotional torture for her.[72]

Later, her "Lines for an Abortionist's Office" was published in W. W. Norton's anthology, *The New Caravan*:

> Close here thine eyes, O State:
> These are thy guests who bring

To gods with appetites grown great
A votive offering.

Know that they dare defy
The words of law and priest—
(Better to let the unborn die
Than starve while others feast.)

The stricken flesh may be
Outraged, and heal; but mind
Pain-sharpened, may yet learn to see
Thee plain, O State. Be blind:

Accept love's fruit: be sleek
Fat and lip-sealed. (Forget
That Life, avenging pain, will speak!)
Thrust deep the long curette![73]

More than forty years later, Ruth was invited to read for a group of students at Cal State (she doesn't specify which one). She wrote to Jerre Mangione that their reading background in modern poetry was "about minus." She continued that "these gals are all for wimmin's lib," so she listened amusedly to essays in praise of Sylvia Plath and Anne Sexton. She "didn't know how they'd react to a [sic] elderly has been like me. So the first thing I read them was 'Lines for an Abortionist's Office,' a poem in my first book written in 1931." That made her their sister, she wrote and continued, "It's rather pathetic how little they know about the Thirties, and how much they want to listen, discovering how much in common we who were kids in the Thirties have in common with these kids who are in college today."[74] She would probably not be surprised that today, another forty years later, the same observations would be appropriate.

5

REGIONALISM AND RADICALISM

The more intense the writer's disgust with the environment from which he sought release, the more ferocious his obsession with its details.

Alfred Kazin, On Native Grounds, *379*

. . . the great age of prosperity ends up in the red.

Maxwell Geismar, Writers in Crisis, *274*

Impotence of the Left

Part of a widespread Red Scare,[1] the Palmer raids of 1919–1920 were aimed at the Industrial Workers of the World (IWW), but many Wobblies, as they were called, were immigrants, and the raids fed a larger anti-immigrant movement. The Immigration Act of 1918 (the Dillingham-Hardwick Act) was meant to remedy what the Wilson administration saw as deficiencies of earlier immigration legislation that made it difficult to prosecute anarchists, labor organizers, communists, war protesters, and other undesirable aliens. The act allowed the deportation of such undesirables using administrative procedures of the Department of Immigration and avoiding due process of law. Under this act Emma Goldman was deported.

These were only two dimensions of a larger xenophobia brought on by the American entry into the First World War. Much of that racism was directed at Italians and Eastern Europeans, many of whom were radicals of various stripes from socialists to anarchists. One notorious example was

DOI: 10.5876/9781646422081.c005

the trial of the Italians Nicola Sacco and Bartolomeo Vanzetti in 1920. They were convicted of the murder of a paymaster and a guard at a shoe store in Braintree, Massachusetts, and sentenced to death.

To the left, this was a clear example of racism. They established legal defense funds and appealed on several grounds: recanted testimony, the confession of another person, and conflicting evidence. The trial judge and the Massachusetts Supreme Court denied all appeals. Felix Frankfurter, later a Supreme Court justice, argued for their innocence.

Sacco and Vanzetti languished in prison while writers, artists, and intellectuals organized a massive resistance to their execution. Literati explored every available avenue for six years while there was hope. Ruth lived that nightmare at *The Nation* for weeks as the execution drew near and shared the agony with Paul.[2]

The electrocution in 1927 stunned liberals around the world and in the United States. It was a clear indication of the lack of power of thinkers of all stripes to influence events.[3] The editor of *The Nation* responded with a powerful editorial suggesting that the names of the fishmonger and cobbler had joined that of the carpenter.[4] A 2007 editorial in the same journal suggested that little has changed since that time regarding terrorism, justice, jingoism, isolationism, and the treatment of immigrants.[5]

The night of the execution, Boston was prepared for an all-out revolt that never happened. Mounted police dispersed a march of 300 unarmed intellectuals. A firsthand account conveys the sense of despair:

> Afterward I talked with some of the people who joined that strange nocturnal march. They knew that it was absolutely futile, that everything had been done that could be done, yet as long as Sacco and Vanzetti were alive they could not sit in a hall and talk and wait; they had to make a last united protest. Then came the execution, the catastrophe that nobody had truly believed could happen. Suddenly they wept or fell silent, they separated, and many of them walked the streets alone, all night. Just as the fight for a common cause had brought the intellectuals together, so the defeat drove them apart, each back into his personal isolation.[6]

More than two decades after the event, Iowa's Josephine Herbst recalled that night when she and her husband, John Herrmann, had disembarked from their sailboat in Portland, Maine:

"You remember how Sacco said, 'Kill me or set me free'?" I said.

"I guess they'll kill him, all right," John said. . . .

Once we visited our hotel room. It was a gloomy vault. . . .

"Let's go back to the Italian's," John said. "He had a good face." We climbed on stools. The Italian gave a pleased little sign of recognition. . . . The radio signaled midnight. . . . Now the voice came on. The prison would have to go dark, the lights out, to kill them. It went dark. Then lights on again. Then dark again. The Italian had been standing still as a statue. Now he took off his apron and hung it on a peg. Rolled down his sleeves over strong thick-muscled hairy arms. He reached up to turn out a light, hesitated, his had still on the switch, looking at us . . . looking me straight in the eyes. His face was tense but calm; one of his eyebrows was nicked with a scar. He spoke in a quiet voice, confidentially. "Electricity. Is that what it's for? Is that the thing to do? Seven years they waited. Not bad men. No. *Good* men."

. . . . a conclusive event had happened. What it meant I couldn't have defined. Looking back from this distance [1960], I might add explanations that would signify. But I don't want to do that. I want to try to keep it the way it was, back there, on the early morning of August 23, 1927, when we walked out on the foggy streets, feeling very cold in our sweaters, and reached out to take one another's hand for the walk back to the hotel. Without saying a word, we both felt it and knew that we felt it; a kind of shuddering premonition of a world to come. But what it was to be we could never have foreseen.[7]

In 1977 Governor Michael Dukakis exonerated Sacco and Vanzetti with a statement that they had been unfairly convicted.

The Communist Party offered an organized alternative to this sense of despair, and the burgeoning Soviet Union offered hope in a time of desperation. But, for writers, "the antonym of 'bourgeois' was not 'worker,' but 'artist,' and the remedy that these writers offered was less frequently to do battle than to make individual escape to Greenwich Village, Paris, or a Land of Poictesme [utopia]—to the Left Bank and not to the Left Wing."[8] Their solution to despair would be personal rather than social.[9] Or, as Alan Wald put it, "the stirring social unity of the Great Depression transformed into the quest for fulfillment of private material wants in the late 1940s."[10]

In his novel *Valley of the Moon*, Jack London has the character of the writer and his wife settle down in Sonoma Valley to become landed proprietors and let the class struggle take care of itself.[11] That progression from grim experience to naturalistic writing to landed proprietor is a good description of the projection of Paul Corey's and Ruth Lechlitner's lives as they would seek personal solutions in sunny California. It is at the same time a progression toward clarity and purpose and a final admission of the frailty of the left in the face of massive and often violent organized opposition.

Ambiance of Marxism

Grassroots indigenous Midwestern radicalism was a foundation for a number of native social and electoral movements that formed a rootstock onto which the Communist Party could try to graft itself. In 1997, looking back on his activity in the Communist Party, Alexander Saxton (1919–2012) wrote that the Chicago branch ran separately from the national party. The organizations that made up the party predated the Communist Party as Wobblies (IWW), Socialists, Knights of Labor, Grangers, and others and would have carried on absent any Communist Party. It was not that the communists infiltrated and took over these groups but that they were the logical, "almost inevitable" forms of political activism.[12]

In a national context, that engrained egalitarian and democratic impetus gave Midwestern writers more than a suggestion of radicalism. As the collapse of American capitalism became more pervasive in the early thirties and the success of the newly formed Soviet Union more certain, many writers, fostered by the Communist Party's John Reed clubs, focused on social consciousness.[13] Enough Americans went abroad to join the Soviet effort to warrant their own newspaper in Moscow.[14]

Most writers were not motivated by any explicit political agenda; they just wanted to write from their own experience, and the experience of people of the Midwest was radical. In addition there was a "general ambiance of Marxism" during the thirties and early forties.[15] The radicalism of Herbst, Conroy, Lechlitner, and Corey sprang from the soil of the Midwest. As Richard Pells put it, "to be a 'communist' or 'socialist' in these years meant very little. One was simply a 'radical,' more or less influenced by Marx."[16]

This was a time before the Great Repression in the United States made even "liberal" a banned word and "radical" a cause for investigation. Being a communist was not against the law. Rather, it offered hope in a time of despair. But while coastal communists might be radical, that did not mean they understood the indigenous radicalism of the Midwest rooted in Jeffersonian democracy and the Bible with injunctions such as Matthew 25, "Whatever you do for the least of these you do for me." The rhetoric of the cooperative movement was blatantly anti-capitalist but neither socialist nor revolutionary but Christian and biblical.[17] But in an age of great disparities, even biblical enjoinders to egalitarianism were radical. Such ideas were beyond the experience of the communists of the Northeast.

Literary historian Douglas Wixson observed that "an anti-ideological current runs through the various radicalisms, as if the Midwestern mind functions according to contingency and not principle. . . . Their legacy was grass-roots democratic expression, a spirit of egalitarianism and individualism-neighborliness that seemed at times at odds with the demand for revolutionary change."[18]

Thus, the Communist Party USA could not comprehend Midwestern radicalism.[19] The editor of a little magazine, The Anvil, Jack Conroy suggested "the worker-editor's task . . . [was] to be a witness to his time, and to record the inner *and* outer struggles."[20]

Of Iowan Josephine Herbst, biographer Diane Johnson wrote, "Looking back at her work and her career, it seems that her motive force, and even her intrinsic subject, was never politics so much as this experience of being a Midwesterner, or rather of escaping from the Midwest, and yearning for the culture of the great world."[21] The same could be said of Corey, who was never very political.[22] Farm fiction, however, he knew, was a label that "could stop any story dead in a magazine slushpile."[23] But the life of Iowa's farms was what Corey knew best and was the subject matter about which he could write most convincingly. That tipped him into the category of a regionalist writer.

In the 1920s and 1930s, communists were centered in New York and used "regional" as a patronizing put-down for such writers as Conroy and Corey.[24] The radicals of the Midwest saw the party's 1935 disbanding of the wide-open John Reed clubs in favor the League of American Writers,[25] which accepted only established writers, as undemocratic and concluded that the party was consolidating power east of the Hudson River with no regard for them. That lent urgency to the efforts of radical writers and publishers in the Midwest to continue the struggle for democracy through their little journals.[26]

THE MIDLAND

There is no doubt that Corey was heavily influenced by Midwestern regionalism centered on Iowa's little journal, *The Midland*.[27] Positioned at the University of Iowa, John T. Frederick founded *The Midland* in 1915. In 1925, small-town Iowa newspaperman Frank Luther Mott joined in the editing with his "passion for the indigenous" and emphasis on accuracy of details of setting that would reveal the life and experience of the place.[28] Milton M. Reigelman remarked, "In *Midland* thinking, a sin equal to sugar-coating or *mis*using our historical materials was not to use them at all—simply to let them be forgotten."[29] The culture of *The Midland* editors was deeply democratic, tolerant of no artificial distinctions or ranking.[30]

There Is No Style, Only Clarity

Literary historian Walter Rideout distinguished between reformers who try to remedy the faults of an existing social and economic system and radicals who argue that the faults are inherent in the system itself and cannot be remedied without cataclysmic change of the entire system, the distinction between late nineteenth-century progressives and early twentieth-century radicals.[31] Authors of radical novels were spurred by the disconnection between the expectations from their upbringing and their experience of life. They objected to the suffering entailed in the economic and social system and advocated the means by which it should be fundamentally changed. It wasn't sufficient to visualize the glories of a future society. Their experience compelled them to record the wretched details of what they had lived and witnessed. Literary realism was their natural medium.[32]

Writing in *New Masses* in 1930, Michael Gold said, "There is no 'style'—there is only clarity, force, truth in writing. If a man has something new to say, as all proletarian writers have, he will learn to say it clearly in time: if he writes long enough."[33] The reportorial journalism of the muckrakers influenced the radical writers of the 1930s in their detailed reproduction of speech and dialogue, descriptions of physical appearance of characters and settings, and attempts to record "all the facts." The description of industrial and manual processes sometimes resulted in flat description.[34]

In *The Great Midland* (1948) Alexander Saxton fictionalized the work processes of railroad yards, "as a precondition of fictionalizing lives lived in and around those processes." He goes on to say that this reportorial function of literature could be duplicated from newspapers, photographs, manuals, and trade journals, but that would not capture the "simmerings of intention and despair."[35] In an earlier work, *Grand Crossing* (1943), he described the processes of work in a railroad switching yard. Readers grasp the roles of each of the workers and how each task fits into the larger process of making up trains from the scattered individual cars and the even larger context of railroads in the economic system. While that might not be of intrinsic interest, it determines the emotional states of the main characters and affects their plans and futures. Paul Corey did much the same in his story "Number Two Head-Saw."

Regionalism

After 1900, publishing centered increasingly on New York, and from that perspective it seemed regional traditions were dying out. For instance, Malcolm Cowley opined that in the 1930s writers expressed no region.[36] But, in spite of New York's insular views, the hinterlands were developing their own regional traditions.

In addition to the mass media of daily, weekly, and monthly journals were many ambitious little magazines that reached smaller audiences, often irregularly. Editors like Jack Conroy, Louis Stoll, John T. Frederick, and Frank Luther Mott worked in their homes; writers provided address lists and peddled issues; printers worked in their garages; readers paid what they could, and writers contributed copy without remuneration. Such magazines were exercises in communitarian production with volunteer labor.[37]

Conroy, Stoll, and Corey formed a kind of mutual support group for little magazines. Corey gives voice to this in a letter to Conroy:

> Somehow or other, Jack, you've got to weld together a group of writers
> in this country who split things open. We've got to present to each other a
> sharp critical side, and like it; and to the outside a solid good opinion of one
> another. If we let ourselves go to just patting each other on the back, we'll
> end in rotting on the inside like "the Great Tradition." Growth and discipline,
> should be a sort of slogan or something.[38]

Little Magazines and Slicks

Access to printing presses and the postal service along with available writers and willing readers facilitated the little magazines and the epistolary relationships among writers and editors via frequent correspondence and sometimes irregular publication. One product of the mass market and mass production that was instrumental in the creation of literature was the availability of the standardized typewriter, which mechanized and sped up the process of writing. Another was the availability of printing presses and supplies. With the introduction of linotype machines in the mid-nineteenth century and their near universal use in the printing and publishing industry by the early twentieth, the old cold-type presses and type were available for the price of scrap. Typewriters and linotype machines made possible the mass publishing industry and the small presses.

Jack Conroy later described the phenomenon:

> Then angry and shabby Jeremiahs in a score of American cities began to
> gather together pennies, cajole printers, commandeer mimeographing
> machines, and to issue publications fired with revolt against a system that
> could permit men, women and children to face starvation in the richest coun-
> try ever inhabited by human beings.[39]

These trends ran against the tide of New York–centered capitalist publishing with its narrow aesthetic standards and constricted entry guarded by agents and editors governed by the blockbuster mentality of mass markets. Jack Conroy and other little magazine editors had freed literature from the market,[40] but a person couldn't make a living in that milieu. One either had to work a day job, as Conroy and Lechlitner and Stoll did, or have the support of a university, as Frederick and Mott did, or not need to make a living.[41] One could do that by locating wealthy patrons, as Herbst did, by being rich in one's own right, or by providing for oneself. That was the position Paul Corey wanted to create on his plot of land on the Hudson River.

"I know how hard it is to hold down a job and write on the side," Corey wrote to Louis Stoll, editor the little magazine *33*:

> That's the reason I gave up the job. It was one or the other. I always do
> all I can to get people to throw up their jobs and write if writing is what
> they want to do. There is one thing I know from my own experience and

that it is, I have done a hundred times better writing now that I'm not tied to a job, than I did along with a job. And if burning candles, or praying or cussing would make 1933 a financial success, so that you could thumb your nose at your factory and boss . . . and turn your time to writing, I'd do all three for you.[42]

It helped that Ruth kept her job to support them both. In contrast, the professional magazines, the slicks, were driven by advertisements; from an economic point of view the texts the writers provided were just the filling between the advertisements, what newspaper journalists called the "news hole." But the slicks could pay, and agents and writers used the little magazines as a way of getting editors of the slicks to notice them. They could hope that H. L. Mencken would spot their work and offer them a place in *The Smart Set*, as he often did.

Wixson marks the end of the period of the little magazines with the demise of *The Midland* in 1933.[43]

Folklore and Fascism

In the English Department at the University of Iowa John T. Frederick carried the regionalist view forward in his little magazine, *The Midland*, until 1933. Neither he nor Iowa's iconic artist, Grant Wood, had anything good to say about any city, especially New York City. Nor about the universal values of literature or art propounded by their colleagues who thought they saw the roots of fascism in regionalism.

Regionalism didn't work the same way in the United States as it did in Europe. By any measure the area of the United States dwarfs any European country. So Germany, for instance, is cheek by jowl with France. Both are small countries. Each insists on its own language and literature. But the United States is made of its various regions. Together they form a country. So regionalists such as Ruth Suckow and Myra Page saw themselves as interconnected parts of a much larger patchwork of regions, not as isolates, each asserting its identity against the others.[44]

The German philosopher Herder argued that a people, a folk, was a part of nature just as the contours of the earth and patterns of weather. The uniqueness of each environment created a distinctive language, culture, and history.[45] In pre-modern Europe each valley had its own dialect, stories, costume, cuisine, and culture that had to be shoehorned into a national template for schools and newspapers so that people would imagine themselves as belonging to a single community, a nation so their leaders could conscript them to fight the working people of other countries.[46]

In the thirties across Europe fascists had a field day with these ideas and began searching the landscape for peasant purity that would be the basis for nationalism in the holy folk, the embodiment of the people writ large. So Gabriele d'Annunzio mined rural lore for vocabulary, topics, and stories to convert into nationalistic poetry, drama, and literature for a fascist Italy.[47]

When American folklorist Constance Rourke started writing about Herder and the American folk, it was a bit unnerving to some on Iowa's side of the Atlantic.[48] Iowa artist Grant Wood alarmed some with his talk of regional art. In 1943 the doyen of art history and Grant Wood's colleague at the University of Iowa, H. W. Janson, wrote that while regionalists wanted to epitomize the qualities of their regions, internationalists knew that great art must transcend region to express universals.[49] In 1946 he explicitly compared Grant Wood and the regionalists to the Nazis.[50]

Fascism can arise in any country, and people were right to worry that it could happen in the United States as well as Germany, Austria, Italy, and Spain. The American South had its regional manifesto in the 1930 collection *I'll Take My Stand: The South and the Agrarian Tradition*.[51] The Southern and Midwestern strands came together in Sinclair Lewis's 1935 treatment of Huey Long, *It Can't Happen Here*, with a Huey Long figure set in New England with his own secret police, called the Minute Men, and concentration camps for recalcitrant journalists. William Randolph Hearst would not have been included in that number because he wrote in 1935, "Whenever you hear a prominent American called a 'Fascist,' you can usually make up your mind that the man is simply a LOYAL CITIZEN WHO STANDS FOR AMERICANISM."[52]

But fascism has some components besides artistic and literary regionalism. In Germany, Italy, and Spain it entailed armed people willing to follow their leader wherever he might lead them. The United States saw one example that some found frightening in Louisiana's populist Huey Long. His rhetoric was familiar to Midwesterners who agreed with his egalitarianism and democracy. His public works, hospitals, and universities were within the realm of the cooperative movement. But he crossed a line and came to resemble the European strongmen when he occupied New Orleans in 1934 with his own personally loyal militia and machine guns in a demonstration of power against the reigning mayor. Only the cool heads of the leaders of the militia and of the New Orleans police force avoided tragedy. In the hot summer of 1935 Huey fell to an assassin's bullet. The popular response to his death showed that his influence among ordinary people was widespread across the United States.[53]

Literature and Culture

While the editors reserved most of the critical commentary in Iowa's pugnaciously regionalist little magazine *Midland* for the Midwest, John T. Frederick reviewed a work by sociologist Howard W. Odum that strove to chronicle the development of the Southern character since the Civil War. Odum collected editorials, pictures, and other materials from newspapers and magazines to illustrate his ideas. Frederick suggested that it would have been more effective to have used the same materials to weave "into an organic fictional structure, based on the career of a single family."[54] Perhaps the empirically minded Odum would have found such a task of fiction writing inconsistent with his mandate as a social scientist, but this is exactly what the expatriate Iowan Paul Corey, not bound by similar strictures, did in his Mantz trilogy.

This speaks to the relationship between culture and literature. Any literature is an artifact of its culture and as such expresses it.[55] This is especially true of regional authors and editors whose purpose was to express their local cultures and distinguish them from those of other locales. Historian and novelist Saxton wrote:

> . . . imaginative writing rests on, or takes off from, a conscious reproduction of human experience. However fanciful, utopian or ultimate the project, it must be put together from segments of everyday living in much the same way that it is composed of words and sentences. And like words and sentences, these ingredients are socially constructed.[56]

Culture of *The Midland*

Milton Reigelman suggests that the stories, editorials, and reviews of *The Midland* express a collective mind-set or culture of the region the Midwest.

These include that the family is a repository of values of the members, that the family is a *part* of the house it inhabits, that the house itself can control the destiny of the family, that the family prevails over its individual members, that the world is stable and coherent, that love is a matter of pragmatic longevity rather than romantic totality, and that the viewpoints of parents and grandparents prevail over those of younger generations. Because there is a premium on mastering a known world rather than charting a new one, older people are more interesting and intelligent than young ones.[57]

Perhaps because it was so buried in the background of the culture as to not be immediately visible, Reigelman did not mention the small and large lies that accompany "Iowa nice" and how they gnaw at the souls of people intent to be straightforward and honest. But it is impossible to maintain a façade of agreeability without such untruths. And the façade is more important than the individual prices it exacts.[58] Iowa small-town native and novelist Stephen Greenleaf wrote, "There are a thousand secrets in a small town, secrets that only stay secret because of the studied and belligerent posture of unawareness that people choose to adopt."[59]

This is just the culture that Paul Corey was fleeing with no sense of nostalgia.

Higher Provincialism

While Grant Wood championed a regionalist approach to art, it was not until 1935 that he responded to Van Doren's revolt against the village in his manifesto, *Revolt against the City*. In the meantime, little magazines such as *The Midland* in Iowa City helped define regionalism.[60] Editor John Frederick, in language similar to that of the German philosopher Herder, argued that different ethnographic, climatic, and geographic conditions produced distinct regional social units.[61] Among these was the Midwest, which exhibited a discrete "regional consciousness" that defined it as a "unit of life in the world," a distinct culture.[62]

By Iowa lights, Malcolm Cowley's condescension was to be expected of New York. The editors of *The Midland* editorialized that New York's publishing, theater, and finance sectors had become standardized with their own comfortable incestuous inner circles. New York was thus anathema to the journal's editorials. "I believe that New York's literary despotism is bad: bad for criticism because New York writers and critics know each other too well and see each other too often; bad for creative writing," Frederick wrote.[63] The remedy was for writers to develop their talent in their own regions free of the debilitating sway of New York. They should remain faithful to the details of their own regions that New York's throttlehold had suppressed. Frederick wrote that writing and living must be "of one piece." Writing should express "the native, natural and genuine, not the remote, external, and artificial."[64]

Frederick had studied English with the University of Iowa's C. F. Ansley, who promoted the Harvard philosopher James Royce's view of "higher provincialism." In his 1902 Phi Beta Kappa address at the University of Iowa, Royce stated that to prevent industrialization from turning Americans into automatons, regions with distinct cultural patterns should encourage their individual natures and promote what he called "higher provincialism." This caused a schism in the University of Iowa's English department between those who agreed with Ansley against those who promoted the study of classical and European literature and denied that the American provinces could produce anything worthy of being called literature.[65]

Ansley was among a group of University of Iowa faculty members who purchased land in the unpopulated clear-cut Upper Peninsula of Michigan in the early 1900s. He took his family to the 200 acres during the summers and began building a cabin. Frederick and his father pooled resources and bought 1,400 acres adjacent to Ansley's place and grew alfalfa, raised cattle, and built a stone house. Visiting writers helped.[66] This was the pattern Corey would follow.

Beyond Iowa

While it is fair to call Corey's writing about Iowa regionalist by any definition, he wrote many other works of fiction that were not rooted in Iowa. Some were set in the western timbering industry ("Number Two Head-Saw," "Green Lumber"); some, in upstate New York (*Shad Haul, Milk Flood*); some, in a vague Middlewest but not identifiably Iowa (*Corngold Farm, Five Acre Hill*); some, in Chicago ("So Democratic"); and some in New York City ("Letter of Reference," "The Cigarette.") Later some of his writing would be set on other planets (*Planet of the Blind*).

It would not be appropriate to view this body of writing from the perspective of regionalism, though there is no doubt that Corey was influenced by his regionalist teachers at the University of Iowa.[67] But throughout the early period of his writing, wherever it was set, Corey focused not on individuals but groups. His stories were often rejected because they lacked a central character, a definite story line, and because they were too long with too many characters and events—too much like real life.

6

NATURALISM AS ETHNOGRAPHY

. . . naturalism has always been divided between those who know its drab
environment from personal experience, to whom writing is always a form of
autobiographical discourse, and those who employ it as a literary idea.

Alfred Kazin, On Native Grounds, *87–88*

. . . I was doing what a writer does, watching and seeing and trying to
understand.

Howard Fast, Being Red, *116*

Ethnography and Naturalism

One way to describe Paul Corey as a writer is as an ethnographic naturalist.
To understand his most expansive writing project, a mural in seven panels
that is the subject of the next chapter, it is helpful to consider the development of ethnography as well as literature and the relations between them.
Ethnography may be the epitome of what American literary historians call
naturalist writing, characterized by objectivity, frankness, amorality. And
pessimism.[1]

For naturalist writers, Philip Rahv wrote in 1949, individuals are determined by their backgrounds, though there is some autonomy, what anthropologists later called "agency."[2] But everything is the product of a social
and historical complex. American naturalism features deterministic forces
manipulating the lives of generally helpless individuals. Sixty years later, literary historian Leonard Cassuto defined American naturalism as writing
that features "deterministic forces manipulating the lives of generally helpless individuals."[3] More poetically, novelist Malcolm Brooks wrote:

DOI: 10.5876/9781646422081.c006

Who would a person love. When would death knock. Mysteries like the wheels of a gear regulated in turn by wars and invasions, earthquakes and famines. In the end the magic of being alive was both created and destroyed by a velocity not perceived but present, each lifetime hurtling toward a light so bright you could but glance before you were forced not only to look away but to forget you ever saw it, for meaning itself was no more than a cipher within that light.[4]

Experience is not unique but typical. Characters are types. The method "of construction is that of accretion and enumeration rather than of analysis or storytelling . . . the massing of detail and specification . . . ," building structures, ". . . out of literal fact and precisely documented circumstance, thus severely limiting the variety of creative means at the disposal of the artist." Such a "quasi-scientific" approach demands neutrality of values.[5]

Trouble in the Sacred Village

With the growth of industrial capitalism came the concentration of people into cities, almost immediately identified as centers of evil as opposed to the more untainted, pure countryside, the locus of all that was antithetic to life in cities. People of villages were pious, patient, caring, and harmonious. People in cities were impious, rushed, annoyed, indifferent, and disputatious. From the early sociological writings about the burgeoning cities and industry, this characterization grew and spread, often in the face of incontrovertible evidence to the contrary.[6]

In anthropology, Robert Redfield gave this proposition, derived from European sociologists, the name of the folk-urban continuum to describe what he perceived during eight months of fieldwork in Mexico in 1926.[7] To the folk pole Redfield assigned values such as isolation, homogeneity, psychological adjustment, piety, cohesion, family, and contentedness; the urban was the breakdown of these. In the US Department of Agriculture's 1940 annual review, Redfield reiterated these notions.[8] This was the sacred village writers were in revolt against, according to Van Doren.[9] So something of the same opposition between the sinful city and the holy village was evident in the world of American letters.

Like Van Doren, Oscar Lewis's study of Tepoztlan in 1943–1944 found political schisms, poverty, maladjustment, violence, disruption, cruelty, disease, and suffering and challenged Redfield's depiction as romantic and hopelessly

Rousseauan. Lewis returned for the summers of 1947 and 1948. In 1949, in an academic paper contrasting the ideals of husband-wife relations in Tepoztlan with the observable behavior of the people, he wrote:

> While some discrepancy between ideal and actual behavior is to be expected in any society the degree of discrepancy in Tepoztlan is striking and reflects the conflict, tension, and maladjustment which characterizes the whole gamut of interpersonal relations there. It is important to note, however, that because of the sanctions in Tepoztlan against the overt expression of aggression in most situations, these conflicts tend to be concealed by a somewhat superficial, conforming and formal type of behavior. It is only when one penetrates deeper that the underlying maladjustments can be appreciated by the investigator. The picture of village life which emerges from our material is therefore quite different from the idealized, almost Rousseauan version of Tepoztlan conveyed by the earlier study of the village by Robert Redfield and later elaborated by [journalist] Stuart Chase. Our data also suggest that perhaps we have had an exaggerated impression of the so-called folk or peasant society as smoothly functioning and well integrated; certainly the quality of human relations as we have seen them in Tepoztlan is hardly superior to that of our urban civilization.[10]

Much the same could be said of Iowa, and Paul Corey said it in his novels that show conflict within families, between families and generations, with government agencies, banks, and other dimensions as well. Some reviewers were shocked to see this exposure of the negative sides of rural life in Iowa, some of which contradicted the carefully cultivated image of "Iowa nice."[11]

Likewise, today agribusiness corporations assiduously promote an image of Iowa farmers as stalwart stewards of the land as they feed the world via sustainable practices while working hard to support their families. Meanwhile, those farmers have polluted the waters of Iowa until they are some of the worst in the nation, have corrupted its politics, have eroded the soils and polluted the air of the state in the practice of a rapacious industrial destruction of the state's natural endowment. But highway billboards and media messages proclaim the image of hard-working Iowa family farmers feeding the world and preserving nature.[12] To be sure, these are carefully orchestrated image campaigns to benefit corporations, but they have their roots in a wider culture. That's why they're effective.

These relationships were just becoming obvious when Paul Corey published his 1948 juvenile novel, *Corn Gold Farm*, in which a heroic high school student promotes conservation measures on his father's farm in spite of all kinds of nefarious opposition from a wealthy neighbor, Harding, who practices destructive farming and holds the mortgage on the loan that was necessary for the protagonist's father to start farming. People in town speculate in farmland and have no interest in saving farms, only in getting all the money out of the land they can get. Tenants want to use conservation practices to build up the land so they can make a living from it. They want to convert from sharecropping to long-term cash leases, but the landlords don't want to risk their income, so they oppose conservation farming. Harding and others apply pressure to fire the high school teacher who had advised the teenager. In this paragraph, deleted from the final version, Corey suggests the processes of disappearing a person:

> Dave Hooker's fight for his job broke in the Alliston Gasette [*sic*]. He was accused of taking advantage of an ex-soldier [the father] who knew nothing about farming to establish his un-American ideas upon the community. Mr. Swanson [who is in league with Harding] said that the ideas he had induced the Blake family to follow had ruined that family. The Voc-Ag program was a failure; its instructor was an undesirable, and both should be forced out of the community.[13]

In the end, though, a drought destroys Harding's crops and livestock and nearly kills his daughter before he comes to the student's farm to beg for water, whereupon he repents his destructive ways. This optimistic ending is not historically warranted. More likely, Harding would have found a way to repossess the farm, as happened in Corey's *Acres of Antaeus* (1946), though there is no documentary evidence that Corey considered such a dire ending.

Anthropologist Walter Goldschmidt argued that the divergence between folk and urban was overstated. It might hold as a contrast between people in Paris or New York and a European peasant community, but urban civilization had penetrated all rural areas of America, "and in more attenuated form has reached every primitive outpost in the world."[14] Goldschmidt was intent to show that American farming was industrialized "throughout the great productive areas in the Middle Western states and the Plains area."[15] With industrialization came urban outlooks, such as the pecuniary market calculus, rather than shared outlooks. The greater the industrialization, the greater the urbanization.[16] He wrote,

> Farm policy has been written for a rural world that is backward, homoge-
> neous, and submissive. The backwardness of the farm population is implied
> by the elaborate structure designed to bring adult education to our rural peo-
> ple. No other element in our population has a government-sponsored pro-
> gram such as the Extension Service which is designed to teach farmers how
> to increase their productiveness and their wives how to make their hats.[17]

By 1940 Goldschmidt saw the trend of developing overarching, secondary organizations such as universities, experiment stations, extension services, the Farm Credit Administration, statistical reporting, rural electrification, and political organizations such as the Farm Bureau. That infrastructure for industrial agriculture was provided by government policies to transform the family farms of Corey's trilogy into industrial farms of *Acres of Antaeus* (1946) or of today.[18] The urbanization of rural America that Goldschmidt reported was part of that process.

These are the struggles that Paul Corey recorded in his novels in personal and particularistic detail. He left the search for universal patterns to other novelists and sociologists just as Iowa's iconic regionalist artist, Grant Wood, left the search for universals to the post impressionists. While Corey falls into the pattern of a regional writer, he never confined himself to his natal Iowa or the Midwest, nor did he exhibit any nostalgia for either farm life or Iowa.

Living and participating in a rural community is experience enough to con-
clude that Lewis's renditions of Tepoztlan are more reliable than Redfield's and Chase's romantic versions. Miner followed his mentor, Redfield, in his tidy ethnographic report that verified a priori notions of purity and pollution rather than challenging them by reporting observable tensions and opposi-
tions. In that respect, Corey's naturalist novels are more reliable guides to understanding farming in the first half of the twentieth century than the ethnographies of some of our most acclaimed anthropologists.

Anthropologists Discover Naturalism

Perhaps the reason anthropologists did not see the Redfield-Lewis debate as part of the larger revolt against the village in American literature that

Van Doren announced in 1922 was that they simply did not read literary critics or historians or much care about American letters. That was for the Department of English, not anthropology. Not until 1973 did anthropologist Clifford Geertz discover what literary people had been calling naturalism ever since before the turn of the twentieth century.[19] He called it "thick description." Then, with the delight of Moliere's M. Jourdain learning, in *The Bourgeois Gentleman*, that he had been speaking in prose all his life, anthropologists discovered that the ethnographic studies they had been writing were—writing. That revelation led them to ineptly embrace the role of literary critic to their colleagues to explain what they had been doing with their typewriters and rebuke them for not doing it better or at least more self-consciously.[20]

Some anthropologists seem to pride themselves in writing mind-numbingly dull ethnographic descriptions with seemingly rote recitations of standardized categories of observations, based perhaps on archaeologists' no less deadening site reports.[21] Turn-of-the-century and early twentieth-century anthropologists had developed speculative theories of the evolution of social orders to explain the rise from simple tribal societies that scholars knew from archaeology to the complexities of the British Empire.[22] These theories would inspire Andrew Carnegie and others to see themselves as the pinnacle of the human species to justify their unapologetic and often brutal rapacity.[23]

The impossibility of developing any evidence for such speculations led some in America to embrace empiricism.[24] This natural history approach to humanity allowed for no such invidious distinctions among the higher and lower, the apex and the base of humanity. All people were equally human; all social orders were equally evolved; no culture was superior to another. The anthropologist's task was to describe the many different cultures like so many butterflies arranged in a case for comparative study by an entomologist.[25] This focus on detailed local description earned these anthropologists the title "particularist."

In his 1922 classic *Argonauts of the Western Pacific*, British anthropologist Bronislaw Malinowski outlined the goals and methods of scientific ethnography. The objectives were to describe the world as it appeared from the perspectives of the people under study and to be able to describe the larger context of their lives, which might not be apparent to the people, what he called a sociological perspective. The point of ethnography was to share peoples' daily experience to appreciate their points of view and to communicate them. American anthropologist A. L. Kroeber wrote in 1922 that this might best be done via fiction.[26]

In the mid-1980s some anthropologists began to agonize over the relationships between writing and realities.[27] Could they really pretend to represent lived realities? What was the status of Malinowski's participant observation? What gives anthropologists the authority to speak for or about other cultures, especially those that were not their own? Where did anyone get the audacity to pretend to represent any reality? Was there any reality?[28]

What they did not bring to the discipline was any less opaque language even though they now spoke in the abstruse words they purloined from an increasingly obfuscatory literary criticism that most anthropologists were willing to leave in the departments of English or literary criticism. It was still as dry as dust and inaccessible to readers unschooled or uninterested in the forensics of French philosophy. But, as Oliver La Farge pointed out, "Many pseudo-intellectuals confuse dullness and lack of clarity with significance. There is even a cult of books which are difficult to read, a sort of pygmy quest of the esoteric, but no sensible person will be fooled."[29] "Good scientists," La Farge wrote, "write simply and clearly, the affectation of an esoteric vocabulary is the hallmark of the second-rater."[30]

Once these anguished anthropologists of the 1980s exposed ethnography as mere writing, it suffered the same fate as fiction in the 1930s—a bifurcation between the dreary and mechanical understanding of reality, as witness for instance, Russell Bernard's *Research Methods in Anthropology* (2011), and the world of private sensibility of the newly revealed postmodernism. As Alfred Kazin wrote in 1942:

> Inevitably, in a world where the public reality can seem so persistently oppressive and meaningless, while the necessity for new means of communication is so pressing, the more sensitive artist is steadily withdrawn into himself, into those reaches of the unconscious where, "one can make a world within a world."[31]

Or as contemporary literary critic Philip Rahv wrote in 1949, these writers wanted to break the novel of its habit of objectivity. "The young men of letters are . . . watching their own image in the mirror and listening to inner promptings. Theirs is a program [of] . . . planned derangement as a means of cracking open the certified structure of reality. . . ."[32] Anthropology was a late-comer to postmodern anguish.

Writing and Reality

No one contests that novelists create fictive worlds. Novelist Stephen Greenleaf wrote that a novelist is, by definition "someone who recklessly disregards the truth."[33] And few ethnographers believe that anthropologists should do either. I think it was John Gardiner who observed that while truth may be stranger than fiction, it is never more dramatic.[34] It's difficult to imagine any drama in most ethnographic writing.

His dedication to naturalism blinded Corey to the requirements of fiction. As Oliver La Farge wrote:

> One of the infuriating things about fiction is that it must always be more credible than fact. We live in a world of coincidences, *dei ex machinai*, odd happenings, and strange behavior on the part of quite ordinary people, but the novelist must make everything seem reasonable and build up to each event, or the reader rejects him. In real life many a situation is completely resolved by a main character's being killed by an automobile, but if we use this device in a story the reader feels flatly cheated.[35]

Critics, agents, editors, and literary historians point out that an author's first obligation is to entertain. A compelling plot line, believable characters, the inescapable logic of improbable situations all contribute. But excessive naturalism is not a means to that end. An ethnographer's first obligation is to faithfully represent observations for other anthropologists. Excessive naturalism. Or, as Greenleaf wrote of the relationship between reality and fiction, "Just because it happened doesn't mean it's good."[36]

And writers of ethnography know that theirs is not a path to literary success—unless they attend to the strictures of fiction.

Alfred Kazin wrote:

> It was not for nothing, then, that the documentary and sociological prose of the depression period often proved more interesting than many of the new social novels. Everyone was writing documentary prose in the depression

period, but novelists were writing it under a psychological handicap. In a period of unparalleled national self-assessment the sociologist writing as a sociologist enjoyed the liberty and revealed the candid freshness of the explorer, the social reporter. He investigated the process of change and described his results; his values were often as obvious as his method, but what he produced had the strength of his knowledge and curiosity. . . . This was the spirit of the new prose writers in the thirties, and it satisfied the country's need to know what had happened to it, a demand that aroused interest in the oppressive *facts* of change, such as migratory workers, silicosis, the emotions of people on relief, and the personality of Adolf Hitler.[37]

Seventy years later, literary historian Alan Wald would write that "one can convert ethnography into fiction,"[38] but he suggests that author Philip Stevenson's fictional treatment of the ethnography of a mid-1930s miners' strike and racial heterogeneity in New Mexico failed as fiction. By his lights, Myra Page may have come closer in her conversion of oral history to fiction.[39]

In one of his novels, anthropologist and novelist Aaron Elkins has two literary people discussing Balzac's *Les Illusions Perdues*:

"It seems to me a marvelous work, full of the most keen observation."

"Of course it is, but an author isn't a sociologist. I don't believe he should be judged on ability to observe, but on the power of his literary style. Balzac's is rudimentary at best, and he's far too melodramatic for my taste, and too moralistic as well."[40]

To escape the confines of ethnographic description, some anthropologists have turned to fiction.[41] Archaeologist, historian, and ethnographer Adolph Bandelier wrote *The Delight Makers* (1890) to make his findings about Pueblo people more accessible.[42] Elsie Clews Parsons's 1922 collection of short fiction, *North American Indian Life*, is a veritable *Who's Who* of American anthropology of the time, including Franz Boas. The illustrations were by C. Grant La Farge, Oliver La Farge's father. In the introduction to that volume, Alfred Kroeber complained that ethnography was often tedious and offered the collection of fiction as an alternative to allow the freedom to explore the thoughts and feelings of other people within the confines of accurate cultural description. Fiction, he wrote, invites psychological treatment, "else the tale would lag."[43]

Oliver La Farge's 1929 *Laughing Boy* won the Pulitzer Prize in 1930 and was made into a movie in 1934.[44] In an introductory note La Farge wrote:

This book is a work of fiction. I have tried to be as true as I knew how to
the general spirit of Navajo things, to customs and character, but all person-
ages and incidents in the story are fictitious. . . . This story is meant neither
to instruct nor to prove a point, but to amuse. It is not propaganda, nor an
indictment of anything. The hostility with which certain of the characters
in it view Americans and the American system is theirs, arising from the
plot, and not the author's. The picture is frankly one-sided. It is also entirely
possible.[45]

In the introduction to a 1951 book about his wife's family, *Behind the
Mountains*, La Farge wrote:

A book of this kind involves the writer in the dangerous, sometimes deadly,
business of "arrangement." The great events of this world are likely to shape
themselves in the natural form of plot, drama, and climax; the small incidents of quiet lives tend to the reverse. It is, I think, no betrayal of truth to
juxtapose incidents in sequences a trifle apter than occur in real life, since no
matter how true the tale, it is worthless if it be not readable.[46]

Another example, praised for its accurate depiction of internal views of
complex events, is Carter Wilson's *A Green Tree and a Dry Tree* (1972).[47] But
neither La Farge nor Wilson was an ethnographer who wrote fiction.[48] Both
were writers who became involved in anthropology and used ethnographic
material to inform their fiction.[49] Edwin Corle was another, though he was
less tied to the requirement for accurate ethnographic description.[50]

Ella Deloria, who studied with Boas, was an ethnographer who wanted
to contribute to social history with her novel of Native American women,
Waterlily (1988), which was published posthumously.[51] Anita Brenner studied
anthropology at Columbia with Boas and got a PhD but never did fieldwork, though she was a prolific and popular journalist.[52] Another Boas
student, Zora Neale Hurston, did fieldwork and presented folklore in fictional form.[53] Biographer Robert E. Hemenway suggests a kind of voca-
tional schizophrenia and an inverse relationship between the quality of her
anthropology and her fiction because "the difference is between the scrupu-
lous reportage appropriate to anthropological description and the unprinci-
pled selectivity characterizing esthetic construction. The reporter describes
as much as she can of the event. The artist uses the event for her own selfish
purposes."[54] In the feminist revival of 1930s women writers, she seems to be

known chiefly as a writer.[55] D'Arcy McNickle worked as an anthropologist and wrote fiction.[56]

An anthropologist who writes both ethnography and fiction under her own name is Jenny White. "Fellow airplane passengers whose eyes glaze over when I tell them I'm a social anthropologist fall right out of their seats with excitement when I mention I've written a novel."[57] Others include Donald C. Wood and Serena Nanda.[58] Aside from Hurston, perhaps the most prolific of these anthropologist-writers was Chad Oliver (1928–1993), known in the anthropology journals as Symmes C. Oliver.[59] In 1995 Walter Goldschmidt (1903–2010) published a short story in an anthropology journal.[60] Kathy Reichs and Aaron Elkins are forensic anthropologists whose fiction has been very successful and translated into television series.

Often mentioned in the same category as anthropological fiction is Laura Bohannan's *Return to Laughter* (1954), published under the pseudonym Elenore Smith Bowen, about her fieldwork in Africa. Another vehicle for ethnography has been memoir, such as Elizabeth Fernia's *Guests of the Sheik* (1965), which conveys the fabric of everyday life among village women in Iraq before Saddam Hussein's assent to power.

Yet another approach is for an anthropologist to translate fiction into English to aid readers in understanding the social backgrounds and emotional responses of the people the anthropologist has lived with and studied. One example is Sidney Mintz's 2002 translation of César Andreu Iglesias's *Los Derrotados* as *The Vanquished*.[61] Sometimes anthropologists mine fiction for insights into the culture that produced it. Examples include Ruth Benedict and Hjorleifur Jonsson.[62]

There are also fictitious works that have been put forward as ethnography. Perhaps the most notorious of these were by Carlos Castaneda, whose books have never been out of print since the University of California Press first started publishing them in the 1960s.[63]

Other fictitious works make use of ethnography. Of these perhaps the best known are Tony Hillerman's detective novels centered on a Navajo reservation. In 1990 the American Anthropological Association awarded Hillerman the Anthropology in Media award. Another example is Mischa Berlinski's *Fieldwork: A Novel* (2007).[64] Some works use an anthropologist as a central character.[65] Examples include Robert Stone's *A Flag for Sunrise* (1981), Amy Waldman's *A Door in the Earth* (2019), Malcolm Brooks's *Painted Horses* (2014),

Susanna Kaysen's *Far Afield* (1990), Lily King's *Euphoria* (2014), Frank Parkin's *Krippendorf's Tribe* (1986), Norman Rush's *Mating* (1991), James Hynes's *Publish and Perish* (1997), and J. Michael Orenduff's mystery novels with his renegade anthropologist.[66] A contemporary writer, Scott Graham, uses an archaeologist as the protagonist in his series of mysteries set in national parks.[67]

Many of the naturalist writers of the 1930s, like the muckrakers before them, had been journalists before they took to fiction.[68] Paul Corey studied journalism at the University of Iowa when it was just beginning to be taught as a trade. It was journalist Stuart Chase who popularized anthropologist Robert Redfield's ethnography of a Mexican village[69] in his comparison of the industrialized machine and time-driven Middletown the Lynds described and the timeless handicraft tradition Redfield venerated.[70]

Almost immediately American anthropologist Ralph Beals chastised both Chase and Redfield for their romanticism, but that rebuke was confined to a footnote in an academic paper in a professional journal.[71] Later another anthropologist, Oscar Lewis, could not find the romanticized Rousseauan peasants of Redfield and Chase but tension, conflict, and maladjustment.[72] Chase's book was a bestseller;[73] Redfield's was not. That's the thing about ethnography. It doesn't sell. Even Lewis's book-length refutation of Redfield could not claim a popular audience.[74]

Oliver La Farge wrote that Lewis went on to experiment and perfect "an important technique for ethnographic reportage: . . . the minute and as nearly as possible total observation of the daily life of single family households. . . ."[75] This is reminiscent of Iowa regionalist John T. Frederick's suggestion to develop a way to weave details of daily lives into an organic structure based on a single family.[76] That is precisely what Lewis did and continued to do in his later work. La Farge goes on to suggest that Lewis handled with a "magnificently brutal frankness" a family that would dismay Chekov and stand Zola's hair on end.[77] Perhaps Lewis could be accused of excessive naturalism, but his work in this vein did claim bestseller status, rare among anthropologists.[78]

Fiction as Ethnography

Naturalist writer Robert Stone may have captured it well when he wrote: "This is the way it is. There is no cure for this. There is only one thing you can do with this. You can transcend it. You can take it and you make it art."[79]

In the story "Number Two Head-Saw," Corey depicts the headsaw as part of a functioning economic unit. He describes how people eat, sleep, talk; how work and the processes of production are organized; and the outlooks of managers and workers. The profit-maximizing employment manager Pendegrass has no qualms denying a meal to a near-starving man or evading corporate responsibility for his death. Workers understand the relationship between food and physically demanding work. The workplace is more than a setting for the story; it determines the story and the relationships among the characters. But more than that, the social organization of the workplace—the relationships among kinds of work, worker's home lives, work and management, and between wages and profits—determines the lives of the workers. In this sense the story is ethnographic.

It could be considered Marxist in the sense that "Marx held that life is not determined by consciousness but consciousness by life, that real life processes and conditions of life determined the inner world of individuals, their beliefs, and their value schemes."[80] Thus, the importance of describing life and work situations and their relationships with beliefs and value schemes, culture.[81] While authors of fiction have the freedom to invent, ethnographers do not.

So it is telling that Corey wasn't content to rely on his memory of a summer's work in a West Coast logging camp a few years before to inform his writing. He wrote to someone intimately familiar with the processes he was describing to check his understandings and descriptions. Corey's letter has not survived, but Frank Nein's handwritten answer has.[82] He confirms the company's bitter opposition to unionism of any kind, "and any talk in Scotia of it will gain your dismissal quicker than anything I know of." He continues, "The criminal syndicalism act still stands; from one to fourteen years in San Quentin for possession of a wobbly [Industrial Workers of the World, IWW] card. There are 14 or 16 men incarcerated today for that offence. But for the past six or seven years it has not been invoked."

Nein details the terms for various parts of the machinery and how they are operated, the titles for the various operators, and their wages as well as the wage cuts prior to Roosevelt's inauguration.

Last October (1933) the Communists invaded Humboldt County with the announced intention of organizing it. In January they pulled out 2 camps of loggers of the Hammond Co. possibly 100 men. It fizzled out for folks up here just can't see Communists. Last May the AF. of L. scattered circulars telling the men they had the right to organize. In June an organizer came in with a charter and is working now in Eureka. Information is hard to get, more or less unreliable when you do get it, but I understand they are concentrating on the woods crews and several hundred men have signed up. . . .

I'm no prophet or son of one but it requires no prescience to predict interesting happenings in Humboldt County in the next six months if the present monstrous strike in San Francisco is successful. It will add immense momentum to the present discontent all over the state.

However my writing time is up so I'll mail this off to you. Sometime before the week is out I'll drop you another page or two and a few pictures I have.

And again let me you give you the assurance I'll try and tell you what little I know about this town and company and believe me the pleasure is all mine.[83]

In his trilogy Corey wished to emulate the work of two anthropologists, Helen and Robert S. Lynd in their Middletown study.[84] Literary historian Douglas Wixson refers to realistic fiction of the thirties about workers as a form of literary ethnography.[85] The Lynds did not mention ethnography as such but wrote that their study was "a pioneer attempt to deal with a sample American community after the manner of social anthropology."[86] Nor does Ruth Benedict mention ethnography in her widely read contemporary *Patterns of Culture*.

Perhaps because he lived the death of family farming, Corey couldn't focus on it, though it is constantly in the peripheral vision of his trilogy. But it was the central focus for anthropologist Walter Goldschmidt. He starts his 1947 book, *As You Sow*, with these words: "From industrialized sowing of the soil is reaped an urbanized rural society."[87] And farming across the United States was industrialized, bringing with it urban values and the associated pecuniary logic. Eastern industry had transformed Jefferson's yeomen farmers into industrial workers who took all the risks of producing raw materials for eastern industry and gained little benefit. Or, as one of Corey's friends and advisors put it, "acquiring capital plant and land on borrowed money; repayment of mortgage by cash crops; sale of crops in world markets . . ."[88]

Ethnography and Farming

Farming was an early venue for American anthropologists. In 1939 rural sociologist Carl C. Taylor, who had been fired from North Carolina's land grant college for being too liberal, was head of the Division of Farm Population and Rural Welfare in the US Department of Agriculture.[89] After meetings with anthropologists and sociologists, he began a series of studies of farm areas. University of Chicago anthropologists Robert Redfield and Lloyd Warner advised.

In the US Department of Agriculture's 1940 yearbook, *Farmers in a Changing World*, is a section devoted to social science that contains a piece by Redfield and Warner, "Cultural Anthropology and Modern Agriculture." Secretary of Agriculture Henry A. Wallace wrote in the preface to that volume that "to build an economic democracy that will match our political democracy, our people must have the facts."[90] The Department of Agriculture at that time contained a Bureau of Agricultural Economics that commissioned studies of agriculture in the West, South, and Middle West. The California study was Walter Goldschmidt's comparison of industrial with family farming, which concluded that industrial agriculture was accompanied by social maladies. Oscar Lewis did a study for the bureau in Texas in 1945 that was published in 1948 when he was an associate professor at Washington University. The title page says he was "Formerly Social Scientist, Bureau of Agricultural Economics, United States Department of Agriculture."

The first of these studies was by Horace Miner in central Iowa. He submitted his report in 1940, but it wasn't published until nine years later, and then by the University of Michigan Museum of Anthropology's Occasional Contributions after Miner had completed his next stint of fieldwork—this time in Timbuctoo.[91] One study, Miner wrote in 1949, caused "a political furore [sic] of such magnitude . . . that the division has been forbidden to publish any more cultural surveys."[92] Miner concluded that "the attempt to make a community study bear upon the implications of a controversial national governmental policy was a tactical error, considering the fact that the study was made for a division of a government bureau involved in the policy."[93]

Ethnographic descriptions are not always popular, especially among the betters of the people they describe. For instance, there was opposition in Oklahoma to John Steinbeck's *Grapes of Wrath*. The better classes in Mexico opposed Lewis's books, though the people they described espoused them.[94]

In 1944 the Bureau of Agricultural Economics commissioned seventy-one county studies like Miner's. Sociologist William H. Friedland sees the demise of the bureau in a study of Coahama County, Mississippi, that found that "race relations between blacks and whites were somewhat less than happy."[95] Representative Jamie Whitten, Democrat from Mississippi, was displeased with this finding and began to dismantle the bureau via his position on the Appropriations Committee. With most of the Mississippi legislators, he voted against the Civil Rights Act in 1957, 1960, 1964, 1965, and 1968.

That bureau was soon disbanded as the emphasis in the land grant universities and USDA shifted from finding out what farmers *were* doing to telling them what they *should* be doing. In 1946 the bureau came under fire "because some of their research was considered objectionable by some members of Congress speaking for powerful external interests," and their funding was cut. In 1953 Dwight W. Eisenhower's right-wing secretary of agriculture, Ezra T. Benson, abolished the bureau.[96] In the 1978 edition of his book, *As You Sow*, Goldschmidt tells how this political movement affected him and his work. For example, his proposal to limit the size of lands people could water from public works came under immediate and incessant attack by large landowners.[97]

With this context of literary and anthropological history, we can better situate Paul Corey's ambitions for a social history of Iowa farming, which he put forward to publishers as his mural in seven panels.

7

THE DEMISE OF IOWA'S FAMILY FARMS

A Mural in Seven Panels

. . . the greatest single fact about our modern writing [is] our writers'
absorption in every last detail of their American world together with their
deep and subtle alienation from it.

    Alfred Kazin, On Native Grounds, *ix*

Three Miles Square

To show publishers his grand vision of the series of novels he wanted to
create, Paul Corey wrote of a mural of the Midwest in seven panels.[1] Of
these seven, he completed four. *Three Miles Square*, the first panel, ready for
publication, he described laconically as "The Mantz Family and their neighbors." This extended quotation is his summary, probably prepared for an
advertising campaign. It gives us his view of what he was doing, his collective, ethnographic treatment of complex human processes. Bobbs-Merrill
published this book in 1939 as the first of the seven-volume series that Corey
envisioned.

> There is nothing nostalgic or sentimental in Paul Corey's realistic and per-
> suasive study of three miles of Iowa farmland and its people. Rather it has
> a Dreiser-like microscopic realism: a down-to-earth unromantic account of
> the life of families who live as close neighbors on a farm checkered area of
> rolling arable land. His Iowans might be any farm community dominated by

DOI: 10.5876/9781646422081.c007

the seasons, surging forward slowly, blessed and harassed by the elements, gossiping, hating, exploiting, fending, loving, cooperating.

Mr. Corey's intimate knowledge of farm life gives authenticity to this work; he has lived the way of life he portrays: its joys and its hardships. When he looks upon cattle grazing on hills that slope green to the rivers; cornstalks turned to a dry death-rattle by drouth; fields stacked with their yellow harvest; furrows steaming in the sun, he views beauty and struggle through eyes of experience. He knows the earth as the farmer knows it—mobile, fertile, generous on the one hand; stubborn, relentless on the other. With brilliance and with detail he makes drama of plowing, dredging, bridge building, early frosts, spring's victorious grip on winter, squabbles over fences, furtive cremation of cholera stricken shotes, auctions, threshing.

"Three Miles Square" [*sic*] is a story without beginning or end: Mr. Corey could extend it either way with equal interest. He begins with the death of Chris Mantz, practical farmer with more than commonplace vision, and follows six years of the family's struggle to carry the tradition Mantz left. The widow, endowed with a good share of energy and mental hardness, and her half-grown son, Andrew[,] live for the pattern Chris drew. Mrs. Mantz with fertile years ahead, denies herself re-marriage, and channels her chocked [*sic*] emotions in religion and the carrying on of her husband's program. Andrew, stubborn, sensitive, lonely, has the dogged determination of a hard-headed youngster to give her wavering moments direction and purpose.

Little human struggles, as much as major crises, give Mr. Corey's canvas the fullness of life: Wolmar Mantz's revolt against Sunday school to assert his individual rights, shocking himself with his purposeless savagery; Otto Mantz's childish emotional confusion when he was caught on the farm of a neighbor whose fields the Mantz's cows had invaded; Verney Mantz, young, independent and full of emotional energy, resentful over Andrew's pre-eminence, slowly changing to warmth when Andrew holds to his own heart the secret of her relations with her lover which he accidently discovered. These people are real; not too pure and not too wicked; not too moral and not too sinful.

The main strength of Mr. Corey's novel is not alone characterization and keen observation of farm life. It has its broader significance: exploitation of farmers by the bankers and politicians; the slow, restless breaking of individualism under the greater force of cooperation. His swift forays out into the country from the focal point—the Mantz family and their farm—thrust below the composite surface to give depth and breadth. The Socialist who came to the community on the dredge, spreading his theories, is not over done; he adds one important detail in a larger, composite canvas.

For Mr. Corey, everything from the earth must be earthy, with fragrance, and surge and bitter struggle. His people are from the earth; his book is from the earth.[2]

The Road Returns

Corey's plan continued with Panel II, which at that time had no title but which Bobbs-Merrill would publish in 1940 as *The Road Returns*. This volume would cover the years between 1917 and the beginning of the entry of the United States into World War I, when Andrew Mantz is conscripted for military service, and continues to 1923. In the first volume, Mrs. Mantz is anxious for her children to improve themselves and move off the farm into respectable middle-class occupations. Andrew becomes heir to his dead father's ambition to someday be an architect, even though that requires education beyond what is locally available. In the second volume, with Andrew gone to the war, Wolmar is responsible for running the farm, but his overwhelming ambition is to be a mechanic, to work with machines, and he bitterly despises farming and rural life and resists his mother's imposition of the management of the farm. Corey described his plan for the second volume:

> This will cover the years 1917 to '23. The land boom hits three miles square, the farmers go speculation crazy; buy and sell, retire and run their farms from the county seat town, and when the crash comes many are left stranded, some with more farms than they can pay for, others with no farms at all and houses in Elm unpaid for.
>
> The Mantz family. Andrew is taken in the first draft, marries a girl at Camp Dix and comes home after the war with the desire to be an architect washed out of him. He settles down to being a carpenter. Wolmar is a poor [bad] farmer and the progress of the farm is slow; he induces his mother to buy unnecessary equipment and when things go wrong, his first impulse is to escape: leave the farm and work in a garage somewhere. Finally, as Wolmar nears his majority, Mrs. Mantz realizes that she must make a change; she feels that Verney and her husband can't be considered to run the farm and Otto

is too young. Moreover, she becomes more determined than ever that Otto shall be educated. She finally gives in to Jensen and sells him the farm.[3]

Jensen is a neighbor who is introduced in the first paragraphs of the first volume. He owns the farm adjoining the Mantz farm, and his ambition is to acquire by one means or another the Mantz farm so he can expand his operation. Throughout the first volume, since the death of Mrs. Mantz's husband, he has been trying to purchase the farm. Andrew refuses, insisting that he can manage the farm, but Jensen tries by underhanded means to induce him to sell. So selling to Jensen seems a defeat for Mrs. Mantz. Corey's plan for the second volume continues:

> —the peak of farm prices has passed and she sells in a falling market. She and the two boys move to Elm, Otto starts to high school and Wolmar goes to work in a garage. They break into town routine and new groups of people are built into the scene. By the end of the book, the post war depression is at its worst, the Rhomer Bank fails, taking with it Mrs. Mantz [sic] surplus cash; farm prices are so low that Jensen can only pay a small part of the interest due [to Mrs. Mantz as she holds the note]; the family has hard sledding. Otto feels the pinch more than Wolmar who is working—and Wolmar has fallen in love.[4]

County Seat

The third volume of the trilogy was published, again by Bobbs Merrill, in 1941 on the brink of the Second World War. Corey's summary:

Book III: County Seat Town

This will cover the years 1923 to '30. I intend to build up the town of Elm in much the same pattern [as] I built up Three Miles square, carrying in through the long slump of deflated farm prices, the spread of the radio, and the hard surfacing of roads, down to the crash which knocked out the whole territory.

Through this the Mantz family will carry on with Mrs. Mantz [sic] determination that in spite of everything Otto will get a university education and rise above his farm background. Wolmar marries, get [sic] into the garage business, repairing end; Otto works in a jewelry store repairing clocks before and after school and here he begins to realize economic injustice when he is paid $3.50 a week for 35 hours of work, and the income to the store from his repair work sometimes amounts to as much as forty dollars. Otto has several adolescent affairs, finally graduates and enters the University. Mrs. Mantz who is left alone, takes in her daughter Vern[e]y, husband and family. She doesn't get on with them, but she puts up with much because of her ambition for Otto. In the spring of 1929 he graduates and she is present at his commencement, which is the high point of her life. She returns to Elm and he goes to Chicago to find work; he is taken on in the business office of a Public Utility firm, but when the crash comes, he, being a late comer, is one of the first to be laid off. He tries to find other work, fails and returns to Elm. In the meantime, Jensen has committed suicide and Mrs. Mantz has the farm back on her hands; she has been renting it for several years and the place is completely down at the heels. Otto's return is a terrible blow to her, because her rugged individualism leads her to believe that anyone with his education can find work if he wants to work, and his inability two find work is only proof that all of her efforts have been wasted on him. Otto realizes the situation, but doesn't know what to do about it, and when he offers to run the farm for his mother she feels that her whole life has been a failure. The details of his taking over the farm are never thrashed out with her because her heart gives out and she dies.[5]

Farm Holiday

Corey projected a fourth volume, to be called *Farm Holiday*, after the Farm Holiday movement. This was never published, but Corey worked elements of it into his *Acres of Antaeus*, which Henry Holt and Company of New York published in 1946.

Book IV: Holiday.
This deals with the farmers [sic] strike, the picketing, the shooting of livestock in freight cars and the raiding of closed banks. Ed Crosby is the farmer leader.[6]

In the first two volumes, Ed Crosby is a neighbor who was friendly with Mrs. Mantz's husband before he died and takes an interest in Andrew's success in running the farm. He has a daughter named Mena, with whom Andrew falls in love, but the romance is disrupted when he is drafted for military service. In the first volume Crosby organizes a cooperative meat ring, a group of farmers who periodically slaughter an animal and share the meat with everyone in the ring; he helps organize a cooperative grain elevator in the second, but the Rhomer bank manages to take over the elevator later before it the bank fails in volume three. Corey's plan for the fourth book continues:

> Otto Mantz and Clipper Jennings are two lieutenants in the fight: Mantz stands for developing a constructive tactic, Jennings for destructive tactics—the two of them are also in love with Crosby's daughter. The strike ends with the government coming to the aid of the farmers, but its tactics have disintegrated them as a body.[7]

Corey wanted to describe the situation of Iowa's farmers dealing with not only the exigencies of farming and weather but also the ups and downs of the market and the incessant pressure from all sides to increase production, expand their operations, and mechanize—to industrialize agriculture at the cost of those who could or did not follow along. An ever-present need for farmers at any scale was to produce sufficient money to pay taxes. Otherwise, even if they did not lose their farms to the bank for lack of a mortgage payment, they could lose them for not having paid their taxes.

During the prosperous years of the First World War, rural counties had built hundreds of new schools and paved miles of roads. There were therefore more public employees and more maintenance costs. Often these construction projects were paid with long-term bonds. By 1930, 60 percent of Iowa's taxes went for roads and schools, and property taxes had doubled in the past fifteen years. When the prices of commodities plummeted, Iowa farmers had to sell eight times as many bushels of corn as they did in 1915 to meet the tax bill. By the winter of 1932, almost half of Iowa farms were tax delinquent and subject to sheriff's sale.[8]

With the 1918 development of lightweight tractors and more efficient farm equipment, farmers sold their draft animals and went into debt to mechanize their operations. The increased efficiency made it possible to cultivate larger tracts of land, and farmers began to borrow money not only for equipment but also to expand their holdings. The more progressive farmers were deeply in debt to finance these improvements. With the horses went a supply of fertilizer and replacement stock. Now farmers had to purchase fuel and replacement parts. In addition, to increase production, farmers began to use hybrid seed instead of saving their own. All of these expenses further embedded farmers in the cash economy so that by the 1930s farmers depended on a steady flow of cash.[9] Farmers couldn't feed their tractors part of their crop.[10]

In February of 1932 a delegation of Iowa farmers from the National Farmers Union returned home with the news that their own President Herbert Hoover thought that emergency assistance from the government would destroy their character and had refused to act. They were angry and determined to do something.[11]

Milo Reno, president of the Iowa Farmers Union, espoused the idea of a farm strike until farmers could get commodity prices in line with their costs of production. Not wishing to endanger the extensive network of cooperatives associated with the IFU, he resigned and became head of a new organization named for the bank holidays.[12]

As early as December, 1931, farmers in at least twelve Iowa counties had massed to halt sales of farm property for unpaid taxes. County officials were elected by their rural base of support, so they made little response for fear of alienating supporters at the ballot box. Besides, farmers were often their neighbors, and they were connected by other ties of community and kinship. By late 1932 farmers had successfully caused tax sales to be postponed.[13]

Early in May 1932 more than 2,000 people assembled in Des Moines to found the National Farm Holiday Association and, having lost all faith in the Hoover administration, decided on direct action in addition to a demand for legislative relief.

Milk producers around Sioux City on Iowa's western edge began blocking all produce except moonshine from entering the city. When truckers tried to break the picket line, people smashed their windshields. Then the picketers resorted to boards studded with spikes to stop trucks. By mid-August about 1,500 people picketed the roads into Sioux City and with farmers in Nebraska and South Dakota closed Sioux City for nearly a month.[14] A settlement increased the price for milk, but not other agricultural products. Local sheriffs managed to break up the strikers in other locations. An armed band attacked a meeting and wounded fourteen near Cherokee in western Iowa.[15]

Much of the action of 1932 to April 1933 was close to where Paul Corey was born, in Marne, Iowa. In Le Mars, Iowa, on April 27, 1933, when District Judge Charles C. Bradley refused to support a mortgage moratorium, 250 farmers invaded his courtroom and nearly lynched him. Governor Herring proclaimed martial law in Plymouth County. The head of the Iowa Bureau of Investigation claimed there was "red backing" for the uprising and there were other charges of communist involvement, but the evidence suggests local leadership. No outsiders were arrested.[16] Such charges suggest an echo of the Red-baiting Farm Bureau.

The attribution of communist involvement was not only incorrect; showing how far removed it was from popular sentiment, the party opposed the Farm Holiday Association and allied groups and denounced the movement and its leaders in its *Producers News*, the party's newspaper aimed at farmers in the area. While there were some communists in the area, the rapidly moving events kept them off balance.[17] When 250 delegates from twenty-six states and thirty-three farm organizations convened in Washington, DC, no more than 25 percent were communists. The meeting condemned any discussion of returning to subsistence farming, peasantry, or abandonment of scientific and technical advances.[18]

No more than 10 percent of Plymouth County farmers participated in the Le Mars event. Nine, all from the corner of the county where the Coreys lived, were later convicted for assault. Most participants were farmers or sons of farmers who were in line to inherit land, but not property owners; those who owned property owned large farms, averaging 230 acres when the average was about 190 acres.[19] Paul Corey later interviewed the lawyer that defended these farmers.[20]

The foreclosures continued, and by 1936 half of Iowa's farmland was operated by tenants.[21] In his fourth volume, Corey hoped to capture some of the drama of these events, many of which had caught the imagination of the eastern press.

Corey's master plan called for three additional volumes to chronicle the demise of Iowa's family farms. The plan continues with Panel V.

"We Are Americans"

While Corey drafted some sketches for this volume, it was never published. The attention of the United States was on the Second World War, not farms and farmers.

Panel V: We Are Americans (Tentative.)
This deals with the county seat town and the effects of the depression upon it. Wolmar is on the verge of losing his garage; he is terribly jealous of

Otto's education and continually belittling him and invariably does the opposite of what Otto suggests, hence he lets the bank have his garage and goes in
for raising bees. He gets involved in some vigilantism, and a brawl with the
unemployed.[22]

To bolster their circulation during the period between the First and Second
World Wars, mass-circulation newspapers replaced their war reporting with
sensational stories of urban crime. There was widespread disregard of the
Volstad Act's prohibition of alcoholic beverages that Corey and his friends
spoofed in their *Vanity Fair* story, "Bullet-Proof Culture."

Dramatic bank holdups contributed to a newspaper-induced myth of a
national crime wave. Iowa's rural law enforcement officers participated in the
fiction. In the face of the illusory crime wave, the rural sheriffs endeavored to
change their image from sedate peacekeeping to militant engagement in a war
on crime. The jail records of the period show that while there was no increase
in rural crime in Iowa, the press treatment of rural crimes changed from being
deplorable anomalies to confirmation for the crime wave. Rural people and
law enforcement officials believed that any lawbreakers must be quickly apprehended and face certain punishment to deter others and stanch the wave.[23]

There was a move to professionalize the sheriffs and integrate them with
state police detectives and the highway patrol. In addition, sheriffs deputized
hundreds of people that bankers had recruited into vigilante bands. When
the New Deal appropriated the myth of the crime wave, Iowa's sheriffs
joined with the rest of the country's constabulary in accepting the leadership
of Washington in a national war on crime.[24]

Local bankers organized vigilante groups, ostensibly to protect their
own deposits. In 1931 the Iowa legislature authorized any sheriff to organize training for vigilantes, and most did. Local Iowa Bankers' Association
representatives lectured the volunteers on the most recent tactics for rural
bank robberies.[25] From the national level, J. Edgar Hoover of the FBI told the
sheriffs "that they were civilization's saviors and guarantors," and the FBI
assumed leadership across the country and provided technical aid.[26] More
than sixty Hollywood movies bolstered the image of the FBI as modern
crime fighters confronting public enemies in a war on crime.[27]

FBI director J. Edgar Hoover consistently and insistently preached that
the salvation of America from an epidemic of lawlessness depended on

its police led by Hoover, and Iowa sheriffs echoed this line.[28] The Iowa Banker's Association got the state to assign a full-time detective to bank robbery investigations, installed the police radio system, and raised the largest vigilante force in the state's history. Local sheriffs deputized them all, and joining a vigilante group became fashionable. In addition the IBA offered free army surplus rifles and pistols to the vigilante groups. The American Legion and National Rifle Association added sawed-off shotguns.[29] The NRA hosted shooting competitions. The primary motive for rural residents joining the vigilantes was "having fun with guns" and getting new guns.[30] During the forty robberies of the 1920s these vigilantes caught 128 bank robbers and killed 9. Then began the decline of the movement until by 1930 most had disbanded as rural banks failed across the state. By 1932 about 300 of Iowa's banks followed Corey's Rohmer bank in closing.[31]

Corey's Iowa panel continued:

> He [Otto] talks up his bee business until so many others out of work take it up that he can't get any money for the honey. He is about at the end of his rope when Otto finally gets him a job as a mechanic for tractor-groups, i.e., several farmers owning a tractor and equipment between them.[32]

A handwritten synopsis of this book, under the title "We Are Americans," outlines Corey's plan:

> Wolmar Mantz wants to be thought as good as the average business man in Elm, including the banker. He hasn't the education or the social position but he tries to justify this lack by comparing the lacks in others. However his position is threatened by the fact that he is just able to pay his taxes and by the fact that he must discuss the mortgage on his property which is falling due. His house is clear and in his wife's name but his planing mill is encumbered. He congratulates himself that he is not as bad off as others.
>
> Lester Stobo (neighbor of Wolmar's) is a college graduate, son of a local grocer who has social standing in Elm. He was hooked by his wife, he finds that clerking in his father's grocery store not fitting the life he thinks a college graduate should have, there is nothing in his environment that interests him—so he drinks to forget and Wolmar Mantz has to sober him up.
>
> James Heflin man about 60 owns two farms which he rents in spite of hard times is able to squeeze enough out of his renters to live. But he finds

time heavy on his hands so he works in Wolmar's planing mill cheap to keep himself occupied.

Mrs. Mantz is not from good social stock, does not have the climbing attitude—is far more sincere and a better business woman than her husband.

Mrs. Stobo is a farmer's daughter who took up nursing to get a husband, but she gets on all right with Mrs. Mantz.

In Chapter II

I should have Tommy say something about Mantz being a good mechanic in spite of his preaching etc.

I need to make a point here of building first the general then the particular. For instance with Laura & Tommy when they first appear I need to build a foundation of high school young folks in general in Elm, then come down to these two specifically. The same with Mantz and the garage business.

Also, first the "general" must be of this town itself, then the shift should be to the homes—general then specific. The same pattern should be applied to their characters. So far I have stuck to close to specific character traits and incidents and the result is then I can't see details.[33]

The following excerpts are from the first chapters of the unpublished "We Are Americans":

Chapter One: As Seen From Above

It was three-forty in the afternoon and the Boeing 247, two hours and forty minutes out of Chicago, roared westward through the torrid sky, altimeter recording a height above three thousand feed and the speed indicator showing one hundred and seventy-two m.p.h. The insulation of the cabin seemed to place distance between the ten passengers and the steady drone of the two motors until their ears no longer heard the sound. In the rear, the stewardess twitched her left wrist and glanced at her watch. "We are now passing over the edge of the drought country," she announced, stressing each syllable, giving it the full benefit of her contralto voice.

Some of the passengers glanced out of the windows and down upon the patchwork of fields that were stubble, pasture and drought-ripening corn, darkened here and there with knots of trees and irregularly traversed by tree-margined, shrunken streams which were as distinct as wrinkles in an upturned palm. On the right side of the cabin, two men looking down, saw a town spread out like a huge maple leaf, its streets like multiple veins extending into and becoming absorbed by the surrounding country, and like a stem, pavement bone-white, a road ran out of the center of the town, due south joining the cross-country Federal highway. The edges of this leaf—this town—seemed to curl a little as if it had begun to dry.

One of these two men, who now looked down upon this town, had shifted his dark eyes, turning his heavy, gunmetal blue features, from a letter half concealed by his pudgy hands which crumpled the upper left and lower right corners hiding most of the printing. But a part of one sentence was revealed: "—and we are expecting you to smash the alien and radical elements in the west-coast unions." Another sentence began: "Your expenses will–" and broke off against the belly of his clutching thumb. The man's glance, slanting through the cabin window, centered upon a new structure—the court house—in the town below.

The other man had turned his head automatically to the window when the stewardess called the attention of the passengers to the area over which they were flying. His sagging cheeks pulled down the outward corners of his eyes until his face had a hound look; the skin of his neck hung loose in soft pink wattles. He peered over the tops of his rimless spectacles down at the town below, tilting his iron grey head and the fingers of his left hand, like five fat shrimps, curled around the large elk's tooth on his watch chain. "Down there," he said with a dry voice directed only for the man behind him, "is where you'll find real Americans still."

The thick hands suddenly folded up the open letter, while the dark face turned sharply, dark eyes searching the canted head of the passenger who had just spoken, over the rattan back of the seat. "Yes sir," he replied as if recognizing the big-boss-look of the man in front, and his heavy cheeks quivered, light shimmering on the gunmetal blue flesh, "guess you're right, sir."

The stewardess said: "In another hour we'll be flying over one of the worst sections of the drought country" She twitched her wrist around again to glance at her watch as if to verify her statement.

A woman on the left side of the cabin, who had been prompted to glance out of the window, turned her face back toward the center aisle, cheeks suddenly pale and slightly greenish, making conspicuous a large brown mole by the lobe of her nostril. She held her left hand tightly over her mouth, while her right made a feeble beckoning gesture. The stewardess got up quickly and took her a paper bag to vomit into.[34]

Chapter Two: The Young Couple On The Town Square

The drone of the afternoon transcontinental transport plane died away with gradually shortening waves of sound, and the young couple standing on the lower side of the town square, below the new court house building, remained silent after their brief greeting as if waiting for the last faint wave of sound from the airplane to pass before they began to talk. They were facing, but not consciously looking, down over the fifteen or twenty cars parked in the cross-hatched center lane of Main Street to the imitation chalet roof on the brick station six blocks away. The expressions of both were constrained as if the youthful fluidity of their face muscles had been parched by the hot wind which lashed about them and rising, boiled in the tops of the trees above the square. On the porcelain pedestal of the drinking fountain beside them was a sign: "Conserve Water."

Without looking at him, she said: "It didn't work either, Tom."[35]

In chapter 9, "Things Beyond Control," we learn that Laura has been trying to terminate an unwanted pregnancy because the outlook for the young man finding employment is impossible, but their attempts on both fronts have failed. Tommy discusses it with his mother:

He [Tommy] turned back into the room. "Mom," he said, "There's somethin' I want to talk to you about." He tipped back his well-shaped head until it found the doorjamb to lean against. "I've got to marry Laura."

His mother laughed. "Ho, Tommy, you and Laura are rushing things. You're both too young to think of getting married. Don't—" She stopped; it took that long for the emptiness in his tone to soak in and make her realize that what he said was not something to be laughed off.

"But I've got to marry her, Mom." His voice was choked with despair and helplessness and his eyes suddenly bright with tears.

She felt no anger at the realization of the situation, remembering suddenly that she herself had "had" to get married. A faint smile began to form around her lips; this blow was lost among the scores of blows the family had received with the past few years. The whole business was so tragic that it seemed almost funny and she began to laugh but shut it off immediately because she knew how close laughing at a moment like this was to hysterics. "Well,—" she began, turning back to the kitchen to hide from him the tears in her own eyes, "I'd always hoped for some grandchildren."[36]

"Tractor"

This work was never published, though Corey used some elements from it in his 1946 *Acres of Antaeus*. He summarized the proposed novel as follows:

Book VI: Tractor.

This is Otto's answer for the farmers. Individual farmers can't afford to buy the expensive equipment and compete with the large land owners and chain farms, therefore their only salvation is for groups of five or six to own mutually this equipment and work it on a co-operative basis.

(The nub of the first third of this novel is expressed in the short story The Green Jackrabbit, carbon of which I am attaching.)[37]

The height of cooperative tractor use in the United States was during the Second World War; in 1943 there were 21,000 such machinery pools in operation. But as members improved their own economic situations, they wanted their own equipment and the groups dissolved.[38]

In the typescript of "We Are Americans" Corey wrote:

Otto laughed. "Fall plowing! The ground's too dry and the horses too feeble. It'd take fifteen head of horses in the condition mine are to pull a sulky in this ground." But his mallet-like head didn't seem to show any seriousness at this black outlook; he looked as if he considered it a joke. He added: "What we need around here is a tractor."

His brother's hand gestured toward the tractor literature on the desk. "I see you've been looking at some 'pro-spectruses.'"

"Yes, there's one there that's just the thing for this kind of a community."

Wolmar looked all around the room and said in a superior tone: "If you got the money to buy it."

Otto was grinning. "If enough of us around here go together, we can buy it. It'll do the work for half-a-dozen farms."

This talk brought Wolmar up short; just another of Ot's cracked brained ideas. "You'll lose out if you try anything like that. You can't make a tractor work on two or three farms at once."

"Who's saying that we'd try to?" Otto had anticipated all this argument with his brother. "We'll take turns. It's got lights and we'll keep the thing going twenty-four hours out of the day—rain or shine. That's the only way we can compete, we small farmers, with the expansion of the chain-farming system up the river. With a tractor among us, we won't have to hire extra help during the year. In a year's time we'll have saved enough to pay our respective shares on the tractor."

Wolmar laughed skeptically. "Just take it from me," he said, "it won't work." He didn't venture any farther into the discussion; that was a defense he'd learned: get into the edge of a subject, then back off with the attitude that he could annihilate his opponent's arguments if he chose.

"How's the garage business?" Otto asked bluntly.

The expression on Wolmar's face changed from an I-can-tell-you-your-tractor-idea's-a-flop to a you're-picking-on-me look. "Pretty slow," he answered slowly; then as rapidly his expression changed to you'll-be-sorry. "I guess maybe I'll let the garage go."

As was intended, this remark shocked Otto and he blurted out: "I'd hang onto it if I were you." In spite of Wolmar's disagreeable nature, Otto always included him in his plans; Wolmar was as disagreeable when he was a kid as he was now and Otto had learned then to overlook it and plan his games to include his older brother. Now the pattern still held. Viewing the farmers' plight, Otto realized that the days of the individual farmer were numbered and the first result of this was the growth of the chain-farm system. Going farther, he realized that unless the small farmer wished to be gobbled up completely, which was an unnatural historical process, he must resist this new movement by developing a movement of his own. To Otto, this movement was a deliberate collectivisation on the part of the small farmers of which his tractor idea was a part. As for small groups of farmers to collectivise their implements they must have a trained mechanic "on call" to render their machines efficient. And Otto had placed Wolmar in this notch in his plan.[39]

Acres of Antaeus

Paul Corey's plan continued to unfold:

Book VII. Chain Farm
  I intend in this book to carry on "Bushel of Wheat; Bushel of Barley" to its logical conclusion.

"Chain Farm" became the 1946 *Acres of Antaeus* from Holt that Corey later considered the fourth of his farm series. While no summary exists in Corey's papers along with the reference to his 1936 novella, "Bushel of Wheat; Bushel of Barley," a contemporary review from WFEA of Manchester, New Hampshire, a CBS affiliate, provides an overview of the novel:

> The latest novel I have read has an interesting title . . . one that catches the eye . . . excites the curiosity . . . Acres of Antaeus, by Paul Corey. From the Jacket design it is obviously a story of farming . . . but . . . who is Antaeus? Well—Mr. Corey explains that. According to Greek legend, Antaeus was a Libyan giant, who, each time he touched the earth, renewed his strength, and thus remained unconquerable. Hercules discovered the secret of his power. . . . lifted him from the ground, and thus overcame him.
>
> In this explanation of the title is found, in brief, an outline of the underlying theme . . . the theory upon which is lain the action of this story about farms and farm empires.
>
> Jim Buckley was working as foreman on a road building gang, saving his money to complete his education at the agricultural college when he met Emily Fletcher at a block dance. Soon they were married, and, shortly afterwards, Jim was jobless. It was during the height of the depression, and work was hard to find. They lived on his savings until they were nearly gone. . . . and then Jim found a job . . . production superintendent for Mid-West Farms.
>
> This outfit was operating farms on the same principles as big business . . . attempting to build an agricultural empire. Jim's job required of him that he follow a production schedule, getting as much work as possible out of the men under him, and, for niggardly wages. It was not a job for a man with a heart or a conscience, and Jim had to forget that he had either. He needed the job, and the money it paid.
>
> Emily did not approve of his position from the first, particularly because her father was in danger of losing his farm to the Corn States Insurance outfit which would, in turn, as soon as they had foreclosed their mortgage, put the lands in the care of the Mid-West outfit.
>
> The farmers who were unfortunate enough to own lands which the outfits wanted, and it was mortgaged, lost their homes . . . and their livelihood. The farms were linked together for mass production, with the company getting as much as possible out of both the lands and the men for as little expenditure of time and money as possible. Acres of Antaeus is a story of unbelievable greed, and of miserable conditions. . . . Men working until they were completely exhausted for a dollar a day. One of the dispossessed farmers,

Clausen, was allowed to remain on his farm to help till the land and oversee the work, receiving the magnificent sum of $20 a month for his labors. Acres of Antaeus, by Paul Corey, is a story of humble people who asked only to be allowed to keep their lands . . . to work . . . and for the right to live decently.

Mr. Corey knows his background. . . . he was born in Iowa himself, and is an enthusiastic farmer. Since his retirement to Putnam County, New York, he has devoted much of his time to writing. . . . This is his fourth novel with a Mid-Western setting.[40]

The Program

Corey continued the description of his program for any agents and publishers he could get to entertain the idea of such a vast undertaking. Behind all of this was his continual production of sketches and short stories, which he submitted individually, hoping to integrate them later into the greater work if he could get some support for the whole plan. He continued the description of his program with a parenthetical statement:

> You will probably wonder at my dabbling with several of these. I don't want to break the program, but if I can't get anyone to carry it through from the beginning, I suppose I'll have to write one of the end ones, and as long as the whole program isn't washed up I can't seem to make up my mind which one of the end ones I should complete. So I've been working on this one and that one and chopping out stories where I can. It's difficult to work in the circumstances, but I believe that if the matter were settled, I could complete the entire program in five years—I'm a hard worker when it comes to writing.
>
> I'm afraid that this summary is too brief and too hasty to give much of an idea of the possibilities of this program, but I think you will understand that I am giving just the barest skeleton, and that in the writing I'll develop the dramatic possibilities a great deal more.[41]

Paul Corey prepared these detailed plans and drafts to outline what he thought of as a social history of the Midwest. Only the first four volumes were published. By that time, the history of publishing and World War II, as well as Corey's conviction that he'd been blacklisted from future publishing, at least on the East Coast, along with the cold winters, motivated Paul and Ruth to consider a move to California.

8

THE ONE OR THE MANY?

Writing Style

It was the abject surrender to naturalism on the part of so many vigorous
young novelists in the thirties that gave the new social novel its basic character.
Alfred Kazin, On Native Grounds, *371*

The Trilogy

Corey's first three books, *Three Miles Square* (1939), *The Road Returns* (1940),
and *County Seat* (1941), center on the Mantz family and compose the Mantz
trilogy. Corey's naturalism shared much with ethnography. It relied on lived
experience for descriptive details, it did not judge anyone, nor did it spare
anyone. It described everyday life in detail, even including minutiae of household arrangement, draft horse technology, and, later, automobile and tractor
technology.

A review of Corey's trilogy in the *New York Times* said, "All three books
have the peculiar absorbing quality of a familiar photographed scene in
which there is an abundance of clearly lighted details in each one of which
one finds a nostalgic pleasure."[1] The nostalgia and surely the pleasure was in
the eye of the reader. Yes, Corey wrote in great detail, almost photographic,
enough to cause the *Times* reviewer to compare him to the super-realists of
the 1920s. One author on regionalism is surprised that some of those details
were negative.[2]

DOI: 10.5876/9781646422081.c008

Corey's novels pile "up details to produce almost a case history of a group representing a particular locale and economic stratum."[3] It's that sense that gives the novels their appearance of being ethnographic, like the descriptions anthropologists develop by the detailed depictions of daily life from sharing the lives of the people of their fieldwork.[4] Rose Feld's contemporary review of one of Corey's novels observed that it "achieves distinction through virtue of its accent on mediocrity; through Corey's deep-rooted respect for the dignity and endeavors of ordinary men."[5] She should have added women.

It would seem that Corey had achieved his goal of a detailed description of the demise of family farms in Iowa and transported his mosaic technique from the short story to the novel. But the novels did not sell as well as Corey or his publisher had hoped. Perhaps with the end of the Civil War in Spain and the secure establishment of fascism across Europe, the purges in the Soviet Union, the German invasion of Czechoslovakia, Poland, the Soviet Union, and France, the Battle of Britain, the New Deal and continuation of the nation's own depression, the American public was not focused on the demise of family farming in Iowa.

It could be that Corey's insistence on group representation, as he thought of it, was not sufficiently captivating to hold the attention of his readers—that they *did* need an individual to focus on.

"Green Lumber"

When he was in college, Paul Corey and twenty others went by train from Iowa City to Scotia, California, where they worked for the summer at the Pacific Lumber Company. Their rail fare, rent for beds in a bunkhouse, and mess hall meals were deducted from their pay, so at the end of the summer Corey returned with the same fifty dollars in his pocket that he'd had when he left Iowa City. In the lumber camp he met refugee Guatemalan revolutionaries and Wobblies who seemed to be decent folks and not to live up to their image as bomb-throwing Bolsheviks.[6] He turned those experiences into a short story titled "Green Lumber," in which he shifted perspectives among a number of individuals, each defined by his or her relationship to the lumber company and its technical processes.

The story opens with the tallyman checking the stacks of green lumber that crews have assembled with one-inch slats between the layers of boards.

The tallyman is thinking about how to get an abortion for the girlfriend he doesn't want to marry. The Italian and the big Swede are stacking lumber. A Portuguese operates the monorail to carry the lumber from the saws to the stacking platform and then to the kilns in the drying yard. The Italian is thinking about his goat that is about to kid and how his wages would help him buy a farm. "Ah, he did not mind being tired when all this great hope was before him."[7] The big Swede, in contrast, was a worker; he had no hope, ambition, or goal beyond the job at hand.

Karl, the spotter who moved the lumber from the monorail down an incline to the drying yard, is thinking that tonight's wages would bring him closer to completing his University of Iowa education to be an engineer. He overhears Reid,[8] another stacker, saying he'd like to point a gun at the president of the lumber company and take his share of the profits. Reid's work partner is worrying about how it will go with his parole officer.

Every worker hears the Swede's bellowing laugh of satisfaction at his work. At the bottom of the incline from the monorail a worker is thinking about Guatemala where he had been the aide-de-camp of the president who had been ousted by the latest revolution. The whistle ends the day's work, and as the workers file out to the bunkhouse and mess hall, we follow their diverse thoughts. The student wants to prove he is adequate—now he is nothing. The president of the lumber company drives by; the employment manager drives by; the "old fashioned girl," the prostitute, greets the men as they file by to the mess hall. The parolee imagines what the employment manager, who is also his parole officer, would say: "The Company's giving you jail birds a big chance. We want to see you make good. Don't try any crooked stuff around here or we'll pack you back to the jug again, pronto!"[9]

Inside his house, we see the discontent of the employment manager's wife as she foils her husband's plot to get her into bed with him. Reid and the parolee take the train to Eureka and plan to quit their work for the lumber company and join a smuggler bringing liquor from Vancouver into the United States. On his way to the speakeasy for a bottle of wine and the company of a couple of girls, the Guatemalan sees the train. After bathing and changing in the bunkhouse, the student struggles with his sense of inadequacy. The tallyman's girlfriend tells him, "You're not ready to settle down and I don't want to make you. I want to do what you want me to do. There

ain't any other way I guess. You . . . fix things . . . and I'll go . . . to Eureka and
do what . . . you say."[10]

As the moon rises, the Italian gets into a shouting match with a fascist
supporter of Mussolini amid a crowd of fellow countrymen; the fascist stabs
the worker to death. The college boy goes to bed lonely. The Guatemalan
gets drunk, and two pretty girls relieve him of his wages. The parolee and
the organizer are on their way out of the story on the train, and the tallyman
decides to marry his girlfriend after all.[11]

Later, as he was sending the story on its rounds of magazine publishers,
Corey wrote to editor Louis Stoll that he had sent it to him "because it's one
of the steps in my approach to writing about groups of people."[12] The reader
sees the intersections of some dozen individuals, each caught in a different
way, in the matrix of the lumber mill. Corey called this kind of story "collective" as versus the traditional story with a proletarian hero that invites the
reader to "impose his or her emotions upon that character [so] the struggle
in the story becomes the personal struggle of the reader," and the reader
cannot develop an understanding of the group.[13]

Corey's Naturalism

To characterize his writing style, Paul Corey suggested his preference for
naturalism and recognized the French novelist Émile Zola as a model. Corey
called himself "the James T. Farrell of farm fiction."[14] Like Farrell, Corey was
persistent.[15] But what could it mean to write like Farrell did? Contemporary
literary critic Alfred Kazin answered the question:

> . . . scene by scene, character by character, Farrell's books are built by force
> rather than imagination, and it is the laboriously contrived solidarity, the
> perfect literalness of each representation, that give his work its density and
> harsh power. As an example for novelists Farrell was as much a blind alley as
> Dreiser, but where Dreiser remained a kind of tribal poet, a barbaric Homer
> who exercised a peculiar influence because of his early isolation and his place
> in the formation of naturalism, Farrell, so much less sentient a mind, grew
> out of the materialism of the early thirties.
>
> [Farrell's] technique was a kind of arithmetical progression. . . . Farrell's
> style was like a pneumatic drill pounding at the mind, stripping off the last

covers of the nervous system. Primitive in their design, his scenes aroused a maximum intensity of repulsion by the sheer pressure of their accumulative weight. If one submitted to that pressure, every other consideration seemed irrelevant or falsely "literary," like Farrell's tone-deafness, or the fact that he improvised his scenes within so narrow a range that the final impression was black unrelieved dullness. This was Life, or at the very least the nerve-jangled and catastrophic life the thirties knew.[16]

Elsewhere, of Farrell's trilogy of Irish immigrant life in Chicago, Kazin wrote:

I had never read an American novel that was such a furious exposé and rejection of the culture that the author himself had grown up in. Farrell had turned himself inside out, had suppressed every personal attachment, in order to paint, in the starkest, blackest, savagest colors, a culture in which there seemed to be no individuals at all; the center was always the group, the gang, the old bunch on the block, and it rested on the most primitive and brutal reactions of blind loyalty to one's own kind and of savage contempt for everything below.

The book was an indictment whose creative force was in the minuteness of detail and in the power of the social logic that he showed in every detail; it was in the pride of assembling these objective social materials and in being able, at last, to bring that dope Studs to the bar of History. . . .

He had turned himself inside out, I thought in wonder as I read his furiously documented book.

The force of the documentation, the unwearing repetition of ugly, crippling details, had an intellectual ecstasy about it. . . . all the lines of force in society pointed to the emptiness and futility of Studs and Studs's kind; they were helpless. The parade on the day of Studs's death, carrying them to judgment, attached all the moral fervor and hope in the world to the author's methods. The facts will make you free. . . . Farrell had drawn up his indictment with absolute faith that all the facts were there, that the facts spoke for themselves, that the facts invited only one possible response. . . . Farrell had given his whole mind and heart to an art based on literal truth and to History as the victory of this truth.[17]

Another contemporary reading was that Farrell "writes well when he is excited or angry, but most of the time he makes his readers trudge through vacant lots in a South Chicago smog."[18] Or, as literary historian Alan Wald put it, "If one lacks prior interest in the material, some of these novels move with the pace of mules across an endless prairie; occasional flashes of power are insufficient to sustain stretches of flat monotony."[19]

Critic and agent Maxim Lieber read Paul Corey's manuscript of *Three Miles Square* and responded that "when the prose travels along at a snail-like pace, with seldom a lift to it, it becomes terribly dull and plowing through a book is truly a labor of love," and offered this advice:

> The very fact that you purpose to follow the pattern of Zola's series shows that you think in terms of naturalism, and while I do not quarrel with Zola . . . some of his books are really magnificent and exciting dramas, the photographic method is indeed out-moded. You introduce for instance, some two hundred odd characters just because they happen to live in your Three Mile Square, but they contribute absolutely nothing to the development of the main story, which is the story of the Mantz family. In my opinion your manuscript would be infinitely improved if you threw out much of the extraneous matter, tightened up the central story and heightened significant incidents. No one will dispute with you that you have splendid material, and I feel that a judicious selection will result in quite a good book. But as it stands now you will have a problem finding a publisher.
>
> I read your own brief remarks at the end of the script, outlining what you propose to do in the way of revision, and I assure you that this is no where [*sic*] near enough if you merely change a comma here and a phrase there, but cut a paragraph here and there, it will simply be playing around. But, what the book calls for is a fearless surgical operation and I hope you will have the courage to perform it.[20]

Paul Rosenfeld of W. W. Norton's annual short story anthology *New Caravan*, which published Corey's "Bushel of Wheat; Bushel of Barley" (the novella that was to become the backbone of *Acres of Antaeus*), wrote:

> I can scarcely advise you what to do with your novel other than take it around yourself to the publishers: I don't think you are the kind of man an agent can successfully handle: you are too original. . . . and you are at liberty, if you desire, to mention my name to them as the source of your direction toward them; and tell them and everybody the Caravan is mighty pleased to be publishing your story-sequence. Whether this will procure the careful

attention you work deserves, I cannot foretell. But I don't think you are going
to have unlimited trouble: your work has too much character and freshness
to pass unnoticed, even in these distraught times; and I am sure you will
make your mark most speedily.[21]

The magazine *Signature* published two excerpts from the trilogy that
came to the attention of William James Fadiman of Metro-Goldwin-Mayer
Pictures.[22] Fadiman wrote that he wanted to see the entire manuscript.[23] He
returned it to Corey, saying that

> the story-line lacks what cinema experts like to call "dramatic high-lights."
> Actually, the simple, honest, earthy lives of your characters contain more
> drama than can be found in a tabloid news-paper, but it is not the type of
> drama that the motion picture world classes as entertainment. I honestly
> believe that your book is a distinguished contribution to farm literature,
> having much of the strength and vigor that is inherent in the soil from which
> your characters draw their vices and virtues.[24]

The editors of *Story* magazine rejected Corey's offer of excerpts because
they thought there had been too many books of the same genre lately.[25] W. W.
Norton held the manuscript during November and December of 1936 before
rejecting it with an invitation to discuss future work with Mr. Norton.[26] An
editor at William Morrow wrote that she had read Corey's story "There's
Always an Accident" in *Scribner's* and invited Corey to let her see the group
of novels he was working on when they were ready. She also encouraged
him to share his writing program with her—how many novels and the prob-
able length of each.[27]

Meanwhile, Corey's friends were doing what they could. In January 1937
Jerre Mangione sent the manuscript to Thomas Y. Crowell Company.[28]

Jack Conroy rejected one of Corey's stories because it lacked "the militant affirmation I'm seeking." Corey answered, "I don't belong to any
political party; I'm an observer."[29] Later Conroy wrote, "What I'd like is
material about workers who are militant and fighting against the depression [and] the system which has brought it on."[30] Repeating what he'd earlier written to Stoll, Corey wrote to Conroy, "I have considered making a
novel of sketches about a plot of Iowa country three miles square, taking
[the] period from 1900 to 1930."[31] None of the New York publishers would
have his *Three Miles Square*, though it interested them. Jerre Mangione

recommended it to the New York representative of the Indiana publishing company Bobbs-Merrill.[32]

Critiques

Corey was submitting the manuscript of *Three Miles Square* to publishers by 1935 when fellow Iowan Buel Beems advised him to rework the book and offered him chapter by chapter advice based on his knowledge of farming in Iowa. He observes that it is the only American novel about cornbelt farming that is occupational: "The most distinguishing feature about the novel is its depiction of the farm as a working place—economic functioning. To be distinguished from novels using farm as setting. . . ."[33]

Beems continued his 1935 letter to Corey regarding the Mantz trilogy project:

> Structure: Swell—collective technique thoroughly justifies itself—has movement and direction; better qua novel than any one of your stories qua stories. . . .
>
> Author's Viewpoint: Your viewpoint throughout is objective and impartial but not unsympathetic—absolutely right, except maybe as to Bessie Mantz [the widow]. Viewpoint is also felt to be close to people and setting, which makes for warmth, vigor, interest; also exposes author to left wing attack as being as confined in viewpoint as his characters. Is Crosby your spokesman?
>
> Style: In the main extremely good—even, vigorous, vivid,—you float an enormous mass of circumstantial detail and get away with it—this is one of the novel's strong points technically, along with the structure. . . . technical descriptions of machines. Inclined to think you overdo this. Very decided value in this however as these mechanical details probably help materially in building the extraordinary impression which the book makes of the farm as a working place with implements far more complicated than hoes and rakes. . . .
>
> Prologue: Worst thing in the book. . . . My idea would be a preface in a matter of fact first person tone. . . . steal all the radical thunder by defining carefully your own position toward farm problems. Also possibly an essay one paragraph long explaining the economics of Iowa farming; acquiring capital plant and land on borrowed money; repayment of mortgage by cash crops; sale of crops in world markets, etc. . . .

Farmers: How about picking three or four farmers, representing different attitudes toward farm practices, say Wheeler, Jensen, Crosby, and Jepsen and pay them more attention than others—make these differences in farm practice appear in terms of farms and character is to the point that the reader recognizes their special representative function.

Omissions: What about county agent; local politics; farm in winter in terms of work; bathing on the farm . . . holidays and fairs, taking kids out of school to work.

Misc: Where does Jensen get his money or credit to buy land; where Billie H. money to buy first auto; where Clem money for car; source of Thorne wealth; where does Crosby get his ideas; how much is this Mantz mortgage anyway; for poor people, you have had Mrs. Mantz handling a lot of appetizing food—probably just details adding up of which you were not conscious in aggregate; think you need to explain to reader a little the under-lying economic principles of corn farming.[34]

In 1938 Marshall Best, an editor at Viking, wrote to Corey:

What you have written is certainly not a novel. It can be called a chronicle, though its interest is not primarily narrative. It is rather a series of panels; and no matter how excellent they are as panels, they do not supply the dynamic interest which the reader legitimately demands of fiction. Your remarks about its sociological significance are well made and would carry weight if the book were offered <u>primarily</u> with that in view. In a work of fiction, how-ever, this must be a secondary purpose which is important only if the book fulfills its function of storytelling successfully.

I offer you this rather dogmatic comment against my better judgment, as we usually find it wiser not to explain our decisions to authors. In this case, however, I feel so much admiration for your writing, the luminous quality of the scenes which it evokes, its sincerity and fine feeling that I feel under obligation to explain our decision. We all agree that you have done very much what you said in your letter and have given us in prose what Grant Wood gives in his paintings. Unfortunately, we feel that in transferring from the medium of painting to the medium of words you have not given enough attention to one of the essential differences between writing and painting. One is static and the other must be dynamic.[35]

Hope

In 1938 Lambert Davis, a Bobbs-Merrill editor who had earlier written to Corey to invite submission of his *Three Miles Square* on the strength of Jerre Mangione's recommendation and his own desire to build his "stable" of new American writers, wrote to Corey's agents:

> As I told you [his agent] in conversation, the chief mistake in THREE MILES SQUARE is that the sociological idea carried him away at the expense of the very human tale which is the core of the book. If he let himself loose simply to write about people, without sociological theories of any kind he could turn this book into something very good indeed. The Mantz family is definitely worth writing about, and he should stick to that job.[36]

His agent went on to urge Corey to meet with Davis, as he was interested in Corey as an author for his publishing house.[37]

Corey responded that as it was, the book would be reviewed by Mumford and Hicks, with whom he had been in correspondence, and that "its broader than usual scope will get critical attention. If I reduce it to just the story of the Mantz family—it will be, no matter how well I am able to do it, just another middle-western farm family story; it will be reviewed by second and third string critics and buried in the review-sheets. It will neither sell nor get attention. I'm basing these conclusions on the reception of middle-western farm fiction for the past ten years."[38]

His agent responded by reiterating her advice to talk with Davis and set up the appointment.[39]

Collective Stories

Corey was publishing sketches and short stories in a dozen magazines—from *Scribner's* to Jack Conroy's *Blast*.[40] Many of them were components of his greater scheme of a social history of the demise of family farms in the Midwest.[41] Thus he developed what John Flanagan called a "lateral" technique of showing the main characters from the points of view of many other people and events to develop a panorama of the area of the three square miles of Iowa where Corey grew up.[42]

In a 1934 letter a British editor wrote that he would like to consider one of Corey's pieces for a series of twelve new stories he was editing for the *Pictorial*

Review, to be called "Story of the Month," and gave the Harvard Club in New York as his temporary US address. In his response, Corey explained his "collective" technique as a term used by other editors to describe his fiction. His criticism of proletarian fiction was that it defeated its main purpose—to generate class consciousness:

> Stories which deal with a single character encourage the reader to super-impose his or her emotions upon that character and the struggle in the story becomes the <u>personal</u> struggle of the reader. And it seems to me that in that type of story the reader can get no real group understanding.
>
> The thing I've been striving for in my recent work is to give the reader the complete view of a group of human beings reacting to a social issue. The plan of the enclosed story was to take five men, each with their personal sex and economic problems and to bring them together over an incident of class injustice; then record the resulting change in their points of view. I'm aware that this brief pattern sounds frightfully mechanical and scientific; but I guess I'm a staunch disciple of Zola.[43]

He mentioned two stories, "Their Forefathers Were Presidents" (in *Story*) and "Nine Pennies" (in Jack Conroy's *Blast*), in which he had tried to create a feeling of the group rather than the individual. When Conroy wrote to accept "Nine Pennies" for *Blast*, he added:

> I saw your "collective story" in the current Year Magazine, and frankly was let down by it. You had great material there—too much of it for a short story—but the method, or so I thought, didn't allow it to come out. Although it was an interesting try for a collective form. . . . Maybe the trouble was you <u>didn't</u> handle the workers collectively, you took a group and examined it one by one, so that what resulted was a series of outlines for proletarian short stories of the usual type, dealing with individuals. (Don't ask me how else it could've been done!) All I mean to say is that the piece gave me no sense of the collective, the unified whole.[44]

About a year later, Corey submitted a story to Philip Rahv, editor of *Partisan Review: A Bi-Monthly of Revolutionary Literature*, which published critical discussions of proletarian literature and reported speeches and discussions of the Communist Party–sponsored Writers' Congress.[45] In his letter of submission for his story "So Democratic," Corey wrote:

Here is a "collective" which I've just finished up and have managed to condense to less than three thousand words in length. It's not as strong as some of my longer "collectives" but I think the triteness of the situation more-or-less accounts for that.

However, I do believe that it is a good example of the scope of the "collective" pattern story as compared to the bourgeois pattern; as for instance if this situation had been handled from the principle character point of view instead the point of view of the whole office staff.[46]

This story regarding the social relations in the office of a real estate publication in Chicago illustrates what Corey had in mind:

So Democratic

The offices of the Harold Compton Publications Inc. occupied the twenty-first and twenty-second floors of the Butte Building on LaSalle Street. Each floor was divided into compartments for the various trade magazines of the Company. The Real Estate News was cramped into narrow quarters on the lower of the two floors, its two windows overlooking the sweeping grey expanse of west Chicago. Whirling around the exposed corners of the tall building, the wind blowing off Lake Michigan lashed the window panes with fine snow.

It was the morning of the last pay-day before Christmas. Jack Freeman, associate editor of the News, happened to be the first of the Real Estate magazine's staff to arrive at work. He pulled the light-cord over his desk; then hesitated, pressing his fingers against his long face. For a moment the smile-lines swinging outward from his lips bent like a fan of steel springs. He had been doing some figuring this past week in his mind. The doctor had told him that his wife must, absolutely must go to a warmer climate because of her serious heart and bronchial condition. Pulling a yellow scratch pad toward him, he jack-knifed his long legs and sat down. He would put a feeling of finality to his mental figuring by setting the numbers down on paper. How far would his Christmas bonus go toward sending Sylvia to Florida?

The sound of tramping feet in the hall dividing the magazine offices grew louder as the employees of the Company arrived and entered the space of their respective publications. Miss Larkin, secretary for the Real Estate News staff, came in panting. Her stays creaked as she stomped to her desk, and her thick arms swinging upward, tucked at the locks of grey hair fallen loose around her ears. She greeted Mr. Freeman and he replied but did not look up from his figuring. "The I.C. was packed to the doors," she grumbled. "I would think they'd put on a few more cars a morning like this." Bending until her broad back-side bulged up like a mushroom, she unfastened her galoshes. "Such weather," she snorted; then pulling out the bottom drawer of her desk she took up a dust-cloth and carefully dusted her desk. "There was a riot down our way," she said. "Over an eviction."

The two girls who checked transfers against the city lot maps, came in together chattering, just as Miss Larkin said: "People are so unruly these days. Really something should be done." She went on with her careful dusting, humming a tune. This was pay-day and the day the Company paid its yearly bonus to employees. She would take her bonus money right over to the bank at lunch time, storm or no storm. Her fingers searched in her handbag until they felt the familiar cloth-rough back of the green pass-book.

Mr. Casey, the advertising salesman, came in swinging his brief-case. He flopped it upon his desk in the corner exuberantly. "Yeoop, pay-day." He boomed. "And bonus." His breath was rank with the smell of his early morning nip. He spun around and pushed up his derby saying: "More dough for your sock, Miss Larkin."

"Uumph!" grunted the secretary.

The two transfer checkers giggled. They had pulled up their chairs to their desks and were thumbing through the stack of transfer reports. Margaret, a sallow-cheeked girl with a tendency to pimples, leaned over her work and turning her head sidewise began to whisper to her companion. "I saw the cutest wool-crepe dress in the Boston Store last night,—only eight twenty-five. Mother promised me I could use half my bonus to buy it with. If dad could just find a job, I could have all my bonus but it takes all Tony and I make to support the family. But mother promised me half my bonus—" Her water blue eyes flashed around the office. "I'm getting so sick of wearing this brown dress. I've worn it for two months straight and it's getting shiney all up and down the back."

Rose, the other girl, worked on steadily; occasionally her sheep's eyes behind their thick-lensed glasses glanced up and about the office. "I want to take a course in Design down at the University, beginning the first of the year," she said. "My bonus will about pay the fee." In her imagination was a vivid vision of herself setting the mode in fashioning beautiful evening gowns.

Margaret didn't seem to hear Rose's remark. "The nicest fellow took me to Frost Land last night," she interrupted. "Gee, he's cute."

Casey crossed the office and thumped Freeman on the back. The younger man casually covered the figures on the scratch pad with his hand. The salesman said: "There're some monkeys down on Halsey Street we might get a box ad from. Will you go down and see if you can get some sort of a story out of 'em. Then I'll call around and soak 'em for the ad."

"Sure," replied Freeman and took the memo which the salesman pinched in his thick fingers.

"Big day! She, fella," continued Casey. "But for me, it's all mortgaged already. Back rent."

"Don't start any riots over it," remarked Freeman, "Miss Larkin doesn't like unruly people."

Casey roared, and the secretary flashed: "Well, I haven't any sympathy with them. Then never save their money and—"

"Maybe they haven't any to save," cut in Freeman.

The salesman boomed: "Now if I should chuck my old woman, I'd take up with Miss Larkin here in a minute. She's got the old sock full. Full, you bet!"

"Uumph!" grunted the secretary. "I'd have something to say about that." And her hard jaw came up with nutcracker violence.

The girls checking transfers broke off their conversation to listen to the other talk and ducked their heads giggling. Casey sauntered over to their desks and resting his freckle-baked hands on the curve of their chairs, began kidding them.

The door of the office opened quietly and Mr. Falloe, the editor, came in, a set smile on his dark face. He stopped at Freeman's desk. "Morning, Jack," he said in a casual tone and his dark eyes passed quickly over the figuring on the scratch pad. "How's your wife?"

"Not a bit better. I've got to figure some way to send her south. The doctor—" Falloe interrupted: "Too bad." Again his eyes shifted to the scratch pad. "Well—" he began; then he stopped. He had ridden up in the elevator with Higgenson, one of the directors of the Company. He left what he had started to say unsaid and his eyes came up from the scratch pad. "There won't be much news breaking today," he remarked and went on into his private office, a small cubicle partitioned off in the corner of the publication's space.

Miss Larking brought in the morning mail and laid it before him. She said in a hoarse whisper: "I don't see why the Company keeps that fellow Casey on for. He's the foulest minded old soak—"

The set smile on Falloe's face flickered into a reality. "Has Bill Casey been deviling you about your sock again, Miss Larkin?"

"Well, I put my money in the bank and I think that's right."

"Yes," remarked Falloe, "if the bank's sound."

His casualness threw the secretary off her track. She had intended telling him all about the riot down her way, but she only asked: "Any letters, Mr. Falloe?"

"Nothing right away, Miss Larkin."

She went out and he sat staring at the incoming mail laying [*sic*] before him. Well, what was the Company going to do? He let his thoughts weight the hint Higgenson had dropped him as they rode up in the elevator.

Freeman had finished his figuring: his bonus would pay his wife's train fare to Florida and if they gave up their apartment the first of the year and he took a hall bedroom somewhere over near Newberry Park, he could send her enough from his salary to keep her there. He turned to the stack of real estate news notices on the corner of his desk.

"Burrr!" muttered Casey, again at his desk pursing his lips and staring out the window at the driving snow. There wasn't much use in going out today but he'd give it a try after while. And he began going through his file of prospect cards. Another drop of "the drink" wouldn't go bad on a day like this. If some of those real estate brokers would loosen up a little. Suddenly he twisted his fat hand around on his thick red neck and peered over the tops of his gold-rimmed spectacles at Freeman's back. "Supposin' we didn't get our bonus."

The clicking of Miss Larkin's typewriter stopped abruptly and the scratching of pencils ceased. "Ho, Mr. Case," squealed Margaret.

Miss Larkin shifted her broad hips on her chair. "They paid the bonuses last year," she said defensively.

The girl with the thick-lensed glasses said: "We've already had a twenty-five per cent cut in our salaries."

Freeman returned to his writing, his hand scrawling more briskly but with nervous jerking.

"Well," said Casey turning back to his cards. "If we don't get it, my land-lord'll be plenty sore. Plen—ty!"

Mr. Falloe came out of his office and stood by Freeman's desk. "Jack," he began, "that Formosa crowd is reorganizing, I hear. You might drop around after a while and see if you can get a story out of them." He spoke in his usual modulated tone. His eyes glanced at the clock on the pillar above the partitions of the offices. It was ten o'clock; the secretary to the treasurer of the Company usually reached the offices of the Real Estate News with the staff's pay envelopes before ten-thirty. "This afternoon, will be all right." He leaned against the partition wall and kicked his rubber heel idly against the green composition board. With his finger he suddenly marked a cross on the figures on Freeman's scratch pad.

"All right," replied Freeman, "I'll drop around." He watched Falloe's long finger marking the cross.

The editor turned and went into his office, closing the door behind him. The staff settled down to work. Occasionally the two girls checking transfers against maps glanced at their wrist-watches. "Ten-twenty," murmured Rose. The telephone rang in Falloe's office and the staff heard his low even voice talking. He stopped; his chair pushed back from his desk scraped on the floor and he came out of the cubicle. "Miss Larkin," he said, "there's a Directors meeting up in Mr. Rorty's office and the editors of the publications have been called in. I'll be up there." And he walked out of the office of the Real Estate News.

After he had gone, Margaret whispered hoarsely: "Oh, I wish they'd hurry."

"What's your hurry?" asked Casey from the side of his mouth, "You can't spend your million until lunch time anyhow."

Miss Larkin asked Mr. Freeman's advice about a news release. Her question interrupted his thinking: if things worked out he could just manage to send his wife to Florida; if things didn't work out— He told Miss Larkin to hold the release until after the holiday.

It was past ten-thirty. The staff heard voices of men coming down the center hallway of the floor. Mr. Falloe came into the News office. His dark face flushed, looked almost purplish. He walked straight into his office and shut the door. Freeman glanced up at him as he passed but the editor seemed to see no one in the room.

Again the staff settled down to work. Margaret whispered to her companion: "You should see my boy-friend skate." But she glanced at her watch nervously.

Down the line of offices came a stop and go motif of office noises. The sound of typewriters and voices would come to a halt in one of the compartments; then a door would open and close twice and the sound of working in the next space would stop while the noise of working in the former compartment would start up again with renewed energy. The stop and go approached the space occupied by the Real Estate News. The door to the compartment next to the News opened and closed and the sound of working ceased, while the sound of someone speaking in a low confident tone took its place. Again the opening and closing of the hallway door; then the door to the News office opened and in walked Mr. Rorty, the president of the Company.

He hesitated a moment inside the door while the working came to a stop. A look of surprise flashed across the up turned face of every member of the News' staff. What was Mr. Rorty doing here? What had he come for? The president stood large and barrel-chested, his ruddy, well-fed features glowed as his glance swept over the desks and workers. The ruled part of his hair, from which the hair was stroked back like wings of wet feathers, swung like a weather-vane. With confident strides he went to Mr. Falloe's office and opened the door. "Mr. Falloe," he said, "The Compton Company wishes you a Merry Christmas."

Mr. Falloe said: "Thank you, Mr. Rorty."

The president turned to the staff secretary: "Miss Larkin, the Compton Company wishes you a Merry Christmas." He bowed and the pink flesh of his scalp, revealed by the ruled part, glistened in the light from overhead. Then he turned and repeated the same greeting to Freeman. He addressed the two transfer checkers as "Young ladies," and the one with thick-lensed glasses mumbled, "Thank you." Mr. Casey rose to the greeting with salesman's poise: "Thank you, Mr. Rorty, and the same to you." Then the president of the Company, bowing to the staff in general, pulled open the door and went on down the hallway to the next compartment. As if in self-defense the News' staff returned feverishly to work.

After the silence in the space adjoining, then the sound of renewed work, Miss Larkin murmured: "Isn't Mr. Rorty so democratic?"

Mr. Falloe's modulated voice called: "Jack, will you come here a minute?"

Freeman rose unsteadily, his forehead damp. His fingers trembled as he pushed open the door to Falloe's office. "Well—" remarked the editor and left the word hanging in the air.

Licking his dry lips, Freeman stammered: "Our—bonus—"

"We have received our bonus," said Falloe quietly.

The associate editor's long legs collapsed and he slumped down into the curved-armed chair at the end of the desk. Falloe's fixed smile hardened into a frown. "What burns me up," he muttered, "is that Rorty would stoop to such a hypocritical song and dance to beat the help out of fifteen thousand dollars."

"Did you expect more of him that the others?" asked Freeman in an unnatural tone.

Falloe looked up out of the corners of his dark eyes. "Perhaps not, Jack." Then he added: "It's too bad about your plans for your wife."

Freeman pulled himself up out of the chair.

"Maybe I can do something—" Falloe went on. He picked up a rubber band and snapped against the wall in front of his desk.

"I don't want charity," cut in Freeman. "I want what I've earned." He stumbled toward the door.

"But a loan—" insisted Falloe; then: "Jack—if—if you even consider organizing the staff—" He broke off adding: "Just don't let me know anything about it—is all."

Freeman's glance questioned the look on Falloe's face. What did the editor mean? He saw the fixed smile come to life for a brief moment. Freeman left the office, closing the door behind him. He stopped by his desk and his hand slowly tore the sheet from the scratch pad. That was that, but—. Casey looked around frowning: "Well, how about it?"

"We've received our bonus," answered Freeman and the spring-lines spreading fanwise from his lips bent back sardonically.

The salesman looked as if he'd suddenly lost a good prospect. The transfer checker, Margaret, cried: "What a shame! My—my—dress." And buried her face in the fold of her arms on the desk. Rose started straight ahead through her thick-lensed glasses at the frosted glass of the partition, not daring to console the girl beside her.

"Put that in your sock, old lady," growled Casey at the secretary; then in a simpering tone: "Isn't Mr. Rorty so democratic?"

"Well—" began Miss Larkin and her eyes and hand searched feverishly for something unknown on the top of her desk. "Maybe the Company needs the money."

"Hell!" said Casey in disgust, adding: "Guess I might just as well call up my old woman and tell her to get ready to move. Pack up the kids and find an old empty box-car to move into."

Freeman crumpled the sheet of note paper in his fist. His eyes flashed angrily around the office. "There's one thing we can do," he began. "We got to do it if we want to keep on living." The heat of his tone arrested the attention of the others. The check girls looked up; Casey twisted in his chair and even Miss Larkin leaned back and her eyes were held by the grim look on his face. "We got to—" he continued; then stopped. The other members of the staff were waiting for what he had to say; they had not noticed the hallway door open and the treasurer's secretary come in with her oblong box filled with the regular manila envelopes.[47]

Payoff

Corey received a letter dated July 22, 1936, from John J. Trounstine saying that John Lehmann, editor of *New Writing*, had returned the story, which Tournstine was returning to Corey with his thanks for prompt cooperation and informing Corey that they would be assembling material for the third issue of the volume during the coming two or three months and that they'd be glad to see more of Corey's work. But the file does not contain an unambiguous acceptance note or letter. Corey does not appear in Ella Whitehead's index to *New Writing* from 1936 through 1950.[48]

"So Democratic" is an example of the style of writing Corey was developing, a style that he thought of as his invention, if not unique to him, modeled on Zola. Interesting in this story is the buildup to the dramatic moment of organizing potential; the point at which one might believe that most of the staff would join Freeman in organizing to assert their needs and rights against management. Then at the last moment, in the last sentence of the story, that mood is undermined by the arrival of management's payment to the workers. Whether Corey intended it or not, this dynamic exactly matched what happened with farmers when they were just sufficiently moved to begin organizing and making known their disquiet in Le Mars and their blockade around Sioux City, Iowa, and other actions of the Farm Holiday Association. Just at that moment, the federal government started giving farmers money, providing them with immediate relief for their urgent problems and undermining any potential for organizing.[49]

Struggle

Corey submitted three related stories—"His Punishment," "Duck Flying High," and "The Fat Does Not Burn"—to various publishers in the early 1930s. Arnold Gingrich of *Esquire* sent a short note saying, "excellent atmosphere, but over-studied effect, and too potentially propaganda for us. Sorry."[50] The editors of the communist *New Masses* rejected the stories with the comment that they liked best "Bushel of Wheat; Bushel of Barley."[51] At *Scribner's Magazine* one editor liked "Line Fence," and another preferred "Duck Flying High," but they also rejected the trio with the invitation to send more stories.[52] In January of 1934, however, W. W. Norton accepted the stories for the anthology *New Caravan*.[53] One of the editors suggested that "Chain Farm" was not a good title and recommended the alternative of "Charlie Turk," as he was the main character. But, he continued, "I see that 'Charlie Turk' for you is not the center of the matter but merely one of its components, and that you really are writing about a group of people in reference to a piece of land."[54] So this editor recognized Corey's notion of collective fiction. Corey received an advance of $61.47 (about $1,149 in 2020 dollars) and a tentative invitation to a cocktail party at W. W. Norton's home to meet the other contributors to the volume and the press, depending on vagaries of scheduling.[55]

New Masses had rejected the novella "Bushel of Wheat; Bushel of Barley," but after it was published, it caught the attention of Ben Field, who was familiar with the director of the Communist Party's International Publishers. He wrote to Corey in 1936 that he'd just read the novella in the *New Caravan*, liked it for its "left punch," and that Alan Calmer of the press wanted to publish it as a pamphlet and distribute it "all over the country" among farmers to foment revolt.[56] He suggested a meeting and asked for directions to Corey's place.

The novella was about small farmers losing their farms to insurance companies and banks who then farmed the land as large industrial units, chain farms. Until the implementation of the New Deal agricultural policies, this had been a source of widespread discontent and sometimes violence. William Fadiman explored the possibilities for converting the novella into a movie with no success.[57] Corey inquired about his rights in the work and learned that he could use it six months after its publication in *New Caravan* with the acknowledgment that it had first been published by W. W. Norton & Company.[58]

Meridel Le Sueur wrote with news of the struggles of the magazine she was helping to publish and that she was happy that Corey's "novelette" was going to be republished.[59] Corey answered that he thought the problem of sales and business management plagued all of the "left press."[60] He continued, "Just publishing a magazine, a newspaper or a book doesn't mean it will sell—they won't sell themselves—they must be sold!" He urged that left publishers organize themselves and offered suggestions and his help. He concluded, "It seems rather ridiculous that writers have not only to write but also to see about selling their stuff to the public, but it seems to me that individual publishing enterprises have let the writers down, and therefore whether we want to or not, we got to do something about it."[61] He went on to say that this was personally motivated because International Publishers was going to reprint his novella.

Almost a year after Field suggested that International Publishers would publish the novella, Corey wrote to the Caxton Printing Company in Caldwell, Idaho, to pitch the same project. He mentioned the positive reviews and that International Publishers had offered to reprint it, but he doubted their ability to reach a Midwestern audience.[62] Early in May he wrote to Ben Field to renew his invitation to visit him and Ruth at Cold Spring and to give him driving directions.[63] Field wrote to Corey that he had been trying to get the novella published but that the final decision was up to Alexander Trachtenberg, founder and director of the press, who was abroad.[64]

Toward the end of July Alan Calmer wrote that Trachtenberg had returned to the office and was too busy to consider Corey's project, but that in the past the press had not had very good success with any form of fiction. He concluded, "I might say, however, he was very offended by the phrase in your letter about 'counter-revolutionary attitude', [*sic*] and I also thought it was entirely in bad taste. I suppose you meant your letter as 'agitation,' but it certainly didn't have a positive effect, and therefore was bad agitation."[65] Corey's previous letter has not been preserved, but this letter started a chain of acrimonious missives that suggest Corey's status as an outsider to the Communist Party. For example, his answer to the previous letter starts:

Thanks for your prompt reply. I appreciate your interest and above all your frankness. I have felt that of all people one should be able to express one's self sincerely and frankly with Communists and receive in return sincere and frank replies. In addition I have felt that I can always rely upon Communists for understanding. . . .

. . . I wish, in the words of Stalin: to be an engineer of human behavior, I am not interested in becoming rich in Hollywood . . . I am not interested in becoming "famous"; I am not interested in becoming a celebrity. I am only interested in building the United Front. . . .

Now as to the case stands: my instrument, i.e., the novelette is being permitted to lie idle. I have offered to help build up its distribution, Ben Field, Jerre Mangione, Weldon Kees, Opal Shannon, and others have offered their help. To let this instrument lie idle; to ignore the willingness to work of willing workers at a time when both effective instruments and the work of workers is desperately needed, constitutes to me sabotage. . . .

If Dr. Tractenberg [Corey consistently omitted the *h* of Trachtenberg] is too busy to get at this ms now, why doesn't he delegate the work to you or somebody else?

As I wrote to Hicks some time ago, it seems that the only way to get attention and action out of the left-movement is to suck-up to cliques and bureaucrats. And if one must sacrifice one's social integrity to do that, one might as well suck up to the reactionary crowd and be paid handsomely for it. All of which doesn't mean that I am going to, but if I will not be allowed to work and my stuff will not be allowed to work for the movement; then I shall be forced to turn to other outlets because congenitally I must write. . . .

Now get this straight, Alan, I'm being perfectly objective with you. I have absolute respect for both you and Dr. Tractenberg, but I am at a cross-road where I must either be allowed and encouraged to work toward the United Front, or where I must take the other course. I am not difficult to work with, but I must be allowed to work. And I refused to be "disciplined" in any bureaucratic fashion. If my scripts are off the line I am quite ready to bring them to heel, but I cannot stand having my energy thwarted. Direct it anyway you wish as long as it is writing but let me stand still and ferment for a little while and I'll explode.[66]

Corey went on to say he was writing only to Trachtenberg and Alan and not going to either the communist or other press because he thought such matters should be private. "I don't believe in airing disagreements on policy or tactics to the public, any more than I would criticize the Communist Party to a non-communist." Corey would later have the opportunity to suck up to the reactionary crowd.

A few days later, Ben Field wrote to Corey that he'd spoken with Trachtenberg and received an equivocal response, that Trachtenberg doubted the work would sell as a pamphlet but he would consider it again if Field would assemble a committee of people who would help with sales.[67] Calmer answered Corey's letter:

> Listen, nobody was trying to "discipline" you. How could I, a clerk in an
> office & a non-party member, etc. presume to "discipline" you, who for all I
> know may be a party member? All I was doing was to communicate to you
> the reaction of T to your calling him a counter-revolutionist. If you'd call
> me that, it wouldn't have much of an effect on me, as I wouldn't, in the first
> place, take it in an extremely literal sense; but T, who is all politics, and who
> occupies a CC [Central Control Committee?] position, was (naturally I think)
> highly offended to have someone brand him with what is to him the foulest
> thing he could be called. . . .
>
> As to why the story hasn't been published by International yet and as to
> why it can't be printed by them for some time yet, I think I've already given
> you most of the reasons: before we publish something we have to be sure of
> a market (that is, we must go out and get advance orders—a job which only
> T seems able to do), that there isn't any existing market at all for us in the
> fiction field as there is in other fields, etc.[68]

He went on to confide in Corey that the press's distribution was near bankruptcy and had backed up all operations including publishing leaving them with a "load" of books that had been ready for publication for months, including works by Marx and Engles that they couldn't print because they couldn't sell them. He estimated it would take several months to get distribution straightened out and until then the press couldn't publish anything at all. He suggested that Corey's story wouldn't sell to the average farmer because "stories will be bought only by people interested in literature, not in farming."[69] Furthermore, what Corey said about the movement not publishing his work and giving him an opportunity to function as a writer is what "most every literary writer around the movement (including myself) is saying today."[70]

Corey responded that if he'd known Calmer wasn't a party member, he wouldn't have been so hard on him. "I'm tough to CP members on principle. I feel that they spend too much time alibiing the party's defects rather than correcting them."[71]

The correspondence continued through the summer. Corey wrote that the committee set up to sell the book couldn't move until it was published, but Calmer responded that IP couldn't publish the book without certain sales. Corey continued to recriminate Trachtenberg. In September, he addressed a letter to "Herb," enclosed some of this correspondence, and asked him to pass them on to "Bee" so someone in the CP would be aware of his position in this matter. This is the strongest evidence I have found that Corey was not in fact a member of the party.[72]

Ben Field wrote to Trachtenberg on Corey's behalf, but Trachtenberg hadn't answered him. Field explained that the distribution committee was ready to go, but Field wanted a commitment that IP would publish the novella.[73] Corey responded the next day:

> Mr. Tractenberg's behavior has kept me from writing for three of the past six months. I'm back at work now and I don't propose to be upset again by unreasonable bickerings.
>
> If this business can't be done on a frank businesslike basis, then it will be satisfactory to me if the whole matter is dropped and the manuscript returned. International Publishers have definitely injured the potential value of the novelette, but I can't afford to waste any more time or energy arguing with them about it.[74]

Field asked Trachtenberg for a definite answer about the novella and assured Corey, "I am authorized to write that International Publishers will take the novelette."[75] He wrote again to say he understood why Corey would be disgusted with International Publishing but that he thought Trachtenberg wanted to go through with publishing it now.[76]

When Alfred Hirsch, editor of the *Sunday Worker*, invited Corey to contribute to that publication, Corey responded that he was "pretty well washed up on offering what ability I have to Party controlled publications."[77]

Sol Auerbach wrote to Corey from International Publishers in October of 1937 that they planned to publish the story by January as a sixty-four page pamphlet. He continued that he understood from Ben Field that Corey and others would attempt to distribute the pamphlet among farmers, though, he said, "We have had extremely bad results so far in literature work among this strata of the population, and we hope that you really can enlist certain groups. . . . to cooperate in the distribution of this pamphlet." He asked for a list of groups that would help so they could proceed immediately to distribution once the pamphlet was published.[78]

Ben Field wrote that he had been to the office of International Publishing and was sorry to learn that Corey did not want to proceed with the project, for Trachtenberg wanted to go forward with it. He said Trachtenberg had left instructions to write to Corey, but they hadn't been carried out, so when he returned from a trip, Trachtenberg asked Auerbach, who was in charge of the office while Trachtenberg was gone, to write to Corey. He continued, "Unfortunately, more than half a year has passed since we first approached International to publish the story."[79]

Corey answered Auberch's letter saying he had been snubbed for eight months and that there were no valid justifications.[80] Corey then began shopping the novella around to the United Cannery, Agricultural, Packing Workers of America, to Simon and Schuster, Viking, *Story*, and even Jack Conroy's *New Anvil*, which never paid authors.[81]

Marshall Best of Viking wrote that he "felt the social critic [was] getting the upper hand over the novelist occasionally," though he admired Corey's writing and regretted "that you are not more concerned with the novelist's first business, which is to tell a story by means of characters."[82] In early November 1940 Lambert Davis, the editor from Bobbs-Merrill who had moved to Harcourt, Brace and Company, wrote to thank Paul and Ruth for a recent visit to their place at Cold Spring and to say that Harcourt, Brace would contract for *Chain Farm* as soon as the third volume of the Mantz trilogy, contracted to Bobbs-Merrill, was complete.[83] He reviewed the agreements they had made while discussing the book at Cold Spring, among them that it be "recast from its present revolutionary slant into the social history framework."[84] Then Mavis McIntosh, Corey's agent with McIntosh & Otis, wrote to tell him that the contract with Harcourt-Brace included a payment of $75 (about $1,392 in 2020 dollars) per month for ten months.[85] Davis continued to discuss the book with Corey during visits to Cold Spring.[86]

Every month McIntosh & Otis dutifully sent Corey a check for $67.50, his share of the $75 per month less the commission. Meanwhile, Paul and Ruth's daughter, Anne, was born and the Nazis invaded Europe. Davis wrote that chain farms were increasing all over the country because of the war emergency and inquired about Paul's success with the US Army.[87] For whatever reasons, Harcourt, Brace turned down the much-revised *Chain Farm* to the distress of Davis and McIntosh and argued with Corey's agent that he should return the advance from any royalties for the book he would get from another publisher.[88] Granville Hicks wrote that Macmillan had the *Chain Farm* manuscript but was "getting jittery, like all publishers and anything can happen."[89] Then he wrote that the book never had a chance because "there was too much politics and that the politics wasn't clear," but that like all publishers during the depressed market of the war years, Macmillan was looking for sure-fire bestsellers.[90] After McIntosh & Otis had shopped the book to several other publishers, it finally came out from Henry Holt in 1946 as *Acres of Antaeus*

9

AGENTS AND EDITORS

... to the novelists of the Middle Border writing out in silent bitterness a way of life compounded of drought, domestic hysteria, and twelve-cent corn ... realism ... was the indispensable struggle against the brutality and anarchy of contemporary experience.

Alfred Kazin, On Native Grounds, *10*

Agents

Virginia Rice was Corey's first agent. In 1933, she had to explain to him the terms of the relationship, her commission, that she was to be his sole representative, and that he should not send material directly to editors.[1] She was favorably impressed with his short stories and complemented his writing. She wrote that she didn't understand why they hadn't sold but said she'd try to find markets for them. "Your work is so finished that it is hard for me to believe that you have sold only one or two stories. . . . have you ever sold a story, or have you just had them published in the non-paying experimental magazines? All sales talk, you know. . . . I'm sure you have a novel in you, and the unusual novel always has a better break than the unusual short story."[2]

Corey responded that her enthusiasm helped him out of "a considerable funk." He continued:

Recently I've been turning over an idea for a novel in my mind. It's taking shape. But I'd like to see some of my stories in print first. It's a question of

DOI: 10.5876/9781646422081.c009

morale. I don't feel that I can go at a big job like a novel again until I've had a little attention. And I feel that if I get a little attention I'll produce a novel that'll be something. I know it.[3]

Early the next year, his enthusiasm for his agent, and hers for him, changed dramatically. She responded to his story "Graduation" with "I can't see this for dust," and he riposted:

[That shows] either a shallow mind or shallow reading. No story of mine, the length of Graduation could be read in twenty minutes or half an hour and be grasped. It's impossible for anyone to do it. . . . That's probably the reason my stuff hasn't sold. Editors are accustomed to reading the first and last sentence of each paragraph and half the conversation and if what the author is saying is simple enough they'll get what it's all about. My stuff just can't be read that way. With regard to GRADUATION, it's all there, significance, character and all, and I've said what I intend to say.

My "serious" writing comes from the inside. [D. H.] Lawrence[4] would say, the blood, but I say, the inside, bone, marrow, guts as well. I block out a rough draft in my head; then I make as many drafts as is necessary to shape the story up, after that, the "inside" is through with it. Then it is ready for my three friends . . . [who] read my stuff and make specific criticisms, detailed criticisms.[5]

He explained that his readers were Ruth, Charlton Laird, and Buel Griffith Beems, an Iowa lawyer and writer then living in New York City. If two of the three agree, he would change it; if only one, he would not. "I'm not sentimental about the stuff I write and think [it] . . . must be the word of Jehova[h]."

If you had said the Graduation story wouldn't place for such and such reasons, I would have accepted your word like a lamb. But when you say, "can't see this for dust," you're just one against three, and I must conclude that you have either not read the thing carefully, or your critical faculties aren't so good.

I would like to feel that you are my fourth critic, and that your slant is the selling slant. . . . You have been very generous in taking up with my stuff at all and I appreciate that fact. I am a big risk and you haven't got a penny out of me, but I have never let anybody down yet who didn't let me down first. The thing is we must see eye to eye. Somebody is going to cash in on me in the near future and I'm willing it should be you.

There now, I have known your worst and you've had my blast from the hills. I think we should be able to get together again now.[6]

But it didn't work out that way. Rice responded:

I'm sorry that my unfortunate slang stuck so long in your memory but I must disagree with you one point. I gave plenty of time to the reading of your story. In fact I have spent a good deal of time on your work and, as you say, without profit. I'm sure you will sell eventually but I'm afraid that I cannot gamble with you. I am very, very busy and I couldn't possibly give you the time and attention you seem to feel you need. I'm sorry about this. I really have been interested in your work but you are so sensitive and I am so busy that somehow or other I feel things would not work out satisfactorily in the end.[7]

Four months later Maxim Lieber was representing Corey and continued to do so for a couple of years,[8] but Corey kept corresponding with publishers and magazine editors on his own behalf as well. In 1938, at the recommendation of Jerre Mangione, McIntosh & Otis agreed to represent Corey.[9]

Farm Sketches

Meanwhile Corey was writing his Midwestern farm sketches. The *Anvil* editor Jack Conroy wrote to Corey:

I really shouldn't do it, but I am sorely tempted to hold your "Barn Dance." As a piece of objective realism, it is superb, but if I use it, I'll have to hold it till the third Anvil, due in the Fall. Do you mind? I should like very much to use it.

Your rural stuff is damned good. I seem to see a connection here, as though these sketches were integral parts of a novel or at least a collection of related sketches. Am I right? It seems to me that such a book ought to be a good and timely antidote for the "State Fair" brand of realism. . . .

I hear from Erskine Caldwell that the Summer smut-hounds have put the screws to "God's Little Acre" in New York. That's a shame. This kind of invidious nosing into literary affairs on the part of the preachers and paid reformers ought to be fought vigorously and continuously. And by the way, your own story may cause the ANVIL to be barred from the mails, but we'll risk it. Did you know anything is a damned sight more obscene in a Communist publication than it is in an "arty" Bohemian one? It's a fact.[10]

Corey replied:

> And you're right about those sketches being a part of a novel. A year ago I
> wrote a story Three Miles Square, a series of ten sketches tied together by
> the jackass. That is what Stoll [editor of the journal *33*] is taking of mine only
> he's using them as sketches. I had to combine two for him, and chuck the end
> one, and the Horses and Mules one I sent you he didn't want. So I've been
> doing more of the same thing off and on since, in between my longer stuff.
> I have considered making a novel of sketches about a plot of Iowa country
> three miles square, taking the period from 1900 to 1930. I don't know how it'll
> be and I can't form an opinion until I've done a lot more sketches to see how
> they work in. I think it has possibilities though in spite of Company K.[11] It
> would be much more of a historical and social document as well as fiction.[12]

Later Corey wrote to Conroy:

> I'm glad you liked that thing in The Frontier. I've been considering a novel
> about a similar family. The Windsor Quarterly has just taken a story which
> I may use as one of the chapters, anyway I'll use the family. I'm still toying
> with the idea of giving a picture of the middle-west for about the past thirty
> years; those waves of pseudo prosperity that have always kept the farmers
> hoping and working like bastards and ending always in the banks and insur-
> ance companies getting all their money and land.[13]

So Corey began writing sketches of life in rural Iowa in an area where
people could hear the braying of a single jackass, three miles square. And he
began to find an audience in little magazines.

Erskine Caldwell's *Tobacco Road* about the rural South was spared the
slush pile, Corey concluded, because of its raunchy content. But nothing
happened in that book "that hasn't happened where I grew up in Iowa."[14] To
judge by the reports of rural crime in Iowa between 1920 and 1941, Corey was
not wide of the mark,[15] though the difference between Caldwell and Corey
may be one of manner rather than content. In *Tobacco Road* and *God's Little
Acre* the poverty, ignorance, and sexuality of the characters is evident in every
paragraph; the sexuality broils beneath a thin skin of dubious decorum and
only occasionally breaks through in detailed description, such as this passage
from *God's Little Acre*, one of several. It involves Will and his wife's younger
sister, Darling Jill, but other people are in the flimsy house:

"Take me, Will—I can't wait," she said.

"You and me both," said he.

Will got on his hands and knees and raised Darling Jill's head until he could draw her hair from under her. He lowered her pillow and her long brown hair hung over the bed and almost touched the floor. He looked down and saw that she had raised herself until she was almost touching him.

He awoke to hear Darling Jill screaming in his ear. He did not know how long she had been screaming. He had been oblivious to everything in the complete joy of the moment.

He raised his head after a while and looked into her face. She opened her eyes wide and smiled at him.

"That was wonderful, Will," she whispered. "Do it to me again." . . .

"Will, do it to me again." . . .

Darling Jill raised her arm and rubbed the teethmarks where he had bitten her.

Will wished he could get up and lie on his back, but she still refused to release him.[16]

There are other more vivid encounters, all against the incessant and relentless drumbeat of the father's obsession for his daughters and daughter-in-law.

The subterranean strata of Corey's novels are not throbbing with anything but technical details of hitching teams, plowing fields, threshing grain, and harvesting corn of a people whose natural condition is hard work, hope, and planning, always planning. The sexuality of his characters is apparent but subdued. Their encounters hardly approach the graphic. This is an example from Corey's *Three Miles Square*:

Then came that Saturday night in February after a trying week—she [Miss Larch, the schoolteacher] had been with Billy four times previously and he had been so gentle and attentive—this night was warm with a balmy, pre-spring warmth and in a moment of weakness and self-pitying tears she had let him have his way with her.[17]

Or, from *The Road Returns*, in which Jens and Anita are siblings:

Once they watched a bull serve a cow; then crept silent and breathless up into the hayloft. He had to take a mare to Frazier's jack, when he was about twelve, and as soon as he got home, Anita asked: "What'd he do?" Jens said, "I'll show you."[18]

But even such truncated portrayals of sexuality were more than the subdued and withdrawn culture of Iowa could bear. The difference lies not only in the authors' expression but in the cultures they were describing. One cannot imagine in Iowa, with its disinclination for invidious display, the kind of conspicuous social stratification that could support the notion of the southern belle, much less the expression of simmering sexuality the role requires.[19]

One of Corey's nieces later wrote:

> I had doubts about Paul's rather graphic prose appealing to the genteel sensibilities of most of my acquaintances and in particular to my relatives. My great aunt Rachel in Atlantic wrote me: "So you like *The Road Returns*. I have never read it and am not planning on doing so. I read *Three Miles Square* and that was enough." Some of the family thought Paul must be a communist.[20]

A New Agent

Jack Conroy wrote to Louis Stoll regarding Corey, and Stoll passed on Conroy's remarks:

> Granville Hicks [a New York literary critic] just dropped in on me, and looked over a copy of 1933. He's a scout for McMillan's [sic] and other publishers, and seemed to be impressed with Corey's sketches. I recommended them.[21]

Later Stoll suggested Thomas Uzzell as an agent for Corey and added:

> I have heard from Granville Hicks. He writes that the magazine is lousy . . . and then . . . "I think that Corey and Schorer show up the best in this issue." Corey is still in the "new barbarian" stage that found expression in the Caravan a few years ago, but he has shrewd perception and a certain skill in presentation.[22]

Critique and Response

Stoll sent one of Corey's stories to Uzzell, who wrote to Corey with a detailed critique of the story suggesting he change the shifting points of view to a single one and indicated that he would be willing to read Corey's material and assist in any possible way, "acting both as a literary agent and as a literary critic," in return for 10 percent of the sales. He requested a statement of Corey's literary program so he could predict the volume of his work and asked "the privilege of ceasing my labors with you at any time it should not seem to me to be profitable."[23]

Stoll was happy to have made the connection and wrote to Corey:

Great! Do exactly what Uzzell tells you. Forget about your opinion, you can form it later after you are successful . . . it will then count for more than it does now. Give him what he wants. . . .

Uzzell's letter to me is: "Thanks for sending me the story by Corey. I can't understand why you would turn it down. If I were in your place, I would be tempted to print just about every word this man would write for me. Believe me, he's good. I have asked him to condense somewhat and make a few minor changes, and hope for the privilege of submitting his piece to Harper's."

But for heaven's sake man, don't be too impatient. . . .

But watch Uzzell. He gets hot over you and then all of a sudden drops you like a hot potato . . . keep this to yourself . . . and I'll watch him from my angle and let you know how things stand as you progress.[24]

Corey wrote to Uzzell that he was rewriting the story but said it would be quite a job "because to blast a thing out of the form it's crystalized in and reshape it to a new form is difficult for me. But I'll try to have the new version to you early in Oct."[25]

Corey agreed to the stipulations in Uzzell's letter and outlined his literary program:

Since Jan 1, 1932 I have turned my entire mental energy, such as it is, to writing. I have completed eighteen stories, averaging about 5,000 words. Some of them are in the first draft stage and may never be completed. They may be turned into chapters for a novel. I have the plans for two novels but I refuse to do anything definite with them until I've made some short story collection and have sold something.[26]

Corey and Uzzell engaged in a long and mutually frustrating correspondence. Corey rejects Uzzell's suggestion that they meet until such time as they are in better agreement. "I am slow-witted in conversation; a row of carrots or a pile of wood is a great help to me in testing the validity of a viewpoint." He continued:

I have a strong sense of my own honor and integrity and I feel duty bound to anyone who helps me to realize my stature. Again, I cannot vouch for the future, but I don't believe I have ever let anyone down in the past, who didn't first let me down, thus abrogating my responsibility to them.

There is no danger of me becoming rich. I am already married and my wife is a poet and critic of some recognition. As for throwing up the writing sponge for any other reason, well, I haven't worked for fifteen years to get myself into the position of devoting my whole time to writing, just to turn back at the first black cat that crosses my path. Sickness of course is unpredictable. I have always been physically healthy. I had a nervous breakdown on the Ency Brit [Encyclopædia Britannica], but since then I guard my mental energies pretty closely. That is why I'm in the country.[27]

Having enumerated to his satisfaction the risks Uzzell would take in working with him as a client, Corey went on to list the risks he saw for himself from working with an agent who had been coaching writers for more than fifteen years and never turned out a prominent one. "So far you have only taken a bunch of nimble-witted imitators, clamped them in a mold and turned them out to fill the magazines." Corey wrote that he would be worth investing in and proposes a three-year contract at a higher percentage than the usual 10 percent so that if Uzzell was right in his estimate of Corey's talent, he couldn't lose. "I am not an 'ideal student,'" he continued. "I never have been, there's not enough yes-man about me to make me an ideal student or an ideal underling in any business." He concludes, "Frankly your letters have only irritated me and kept me awake nights."[28]

Later, Corey wrote:

On September 20 I wrote you expressing my satisfaction with your proposition for handling my stuff and giving you an account of my 'literary program'. . . . I'd like very much for you to act as my literary agent. . . .

For the past two weeks I have been turning over in my mind an idea for a novel, but I can't make up my mind to go at it. The idea is pertinent to the present national situation, but the facts that I have sold nothing, have only 'l.[ittle] j.[ournal]' recognition and the question as to how long the present national situation is going to last, makes me doubtful about starting anything sixty or seventy thousand words in length. On top of that, my confidence in my own ability is pretty well shaken at the present time.[29]

Corey continued to send Uzzell stories, and Uzzell provided detailed critique. Rather than rewriting according to Uzzell's advice, Corey defended his writing point by point, sometimes questioning Uzzell's competence as an agent, sometimes his philosophy, and sometimes his taste. At the end of 1933, Uzzell wrote to Corey:

1. Don't use the ad hominem argument on me, please. If we must argue as to whether I know my business or not, please don't dismiss what I say by simply asserting that I am dogmatic or implying that I am not intelligent. . . . Calling names ends arguments without result other than bad feeling.

2. Please don't force upon me a discussion of theory when I raise no such discussion. Such talk is endless and of no profit unless we are collaborating on a treatise. Ask me concrete questions about concrete work you are doing.

3. Please remember that you set the specific problem of getting your work into the established, paying literary magazines. When I offer advice, you reply with a whole flock of complaints and generalizations about writers, editors and art, most of which I agree with, but very little of which is relevant to the job before us. Let's stick to one thing at a time: How to get you into the big time literary books? If you don't want to stick to this and let me see your work and try my suggestions, then tell me and we'll map out another program. How would you like to correspond with me for publication? Seasoned Critic vs. Talented Almost Arrived Author on how writing should be taught! We could make it quite readable.[30]

A Proposal

Corey sent Uzzell a synopsis of a novel, *Acre March*, which later became *Five Acre Hill*. A middle-class couple in their thirties who graduated from college in the mid-1920s have saved their money, spent a year in Europe, and returned in the fall of 1929. March, the husband, gets a job as the illustration director on an encyclopedia, and his wife gets secretarial work. So far, this is straight autobiography. When the encyclopedia is published, both lose their jobs. Living in cramped quarters in New York, "they get on each other's nerves." They want to have a family, but she is getting beyond the age of having children safely. "They are down to the last two weeks of their money. They quarrel perpetually."[31] Prophetically, Corey continued, they see themselves as incompatible and consider separating. A young lawyer friend comes to visit and invites March to lunch at the University Club. March accepts, motivated by the free meal, and in the lounge gets into conversation with a publisher and others. The conversation dwells on the Depression. Most agree that no one could live on the wages of a three-day work week, but March says he could buy an acre of land, build a house, buy a car, and pay for it in two years. The publisher bets him $5,000 against the property he would buy that he couldn't do it. March accepts. The publisher hires him three days a week.

The Marches search for and find some ground, get a loan to purchase it, and start building a one-room house that can serve as the nucleus of a larger one. Corey explains in the synopsis that he will work out in detail of materials and expenses so the book could be a guide. "After all," he wrote, "I know my stuff there because I built my own house." So far the story is still autobiographical with the exception of the wager with the publisher.

March comes to suspect that his wife, Anne, is carrying on with the young garage man on the post road. "He worries and stews." Anne "realizes what he is fretting about and straightens the matter out." Their bond strengthens, and they are "welded together by the mutual struggle." The synopsis continues:

> Then, the second summer, Anne becomes pregnant and they are faced with a catastrophy [sic] If they go on and have the child, March may not win the bet, he'll probably lose his job and they'll be worse off than ever; the cost of an abortion will almost certainly make them unable to win the bet. And in the solution of that situation will be the climax. I want them to decide to have the baby and come through with the bet.

This passage is so closely autobiographical, right down to the ages and characteristics of the couple, the location of the garage, and the unanticipated pregnancy, that it seems as though Corey were looking for a way to imaginatively replay the abortion he had persuaded Ruth to have. Corey proposes to work some of his own philosophy into the story and to start writing immediately and find a publisher by spring or fall at the latest because, "the present economic crisis [1933] . . . will last at least six months and perhaps a year longer and I think that is optimistic."

Corey's synopsis continued:

This idea will find opposition in the proletariat because they'll say the idea plays right into the hands of Capital. And the people who like to work full time and hire the hard work done and play golf once a year for exercise will rage and say it can't be done. But there is a growing class of city worker who wants to get to the country and achieve a certain security, not represented by money and a job, and this class is both among the active and the dreaming types. But again, it may be all rot. I hesitate to go at it because I don't want to waste my time over something as big as it will be that'll be still-born.

I'm keen on the idea as a "way out" sort of thing. I've another idea for a Midwestern novel, that will certainly find favor in the eyes of the proletarian group. It also is a "way out" theme, but I wouldn't undertake it until I had sold something because I'll have to have money to do some necessary research.

What do you think?[32]

It would not be until twelve years later, in 1945, that the book would find a publisher with the excision of the pregnancy and the husband's jealousy. The idea for the Midwestern novel would first become "Bushel of Wheat; Bushel of Barley," a collection of three related stories published as a novella in the 1936 edition of W. W. Norton's *New Caravan* and over the next decade morph into *Acres of Antaeus* (1946). The "guide" dimension of the proposal became *Buy an Acre* (1944).

Parting of Ways

Uzzell and Corey continued their correspondence—Uzzell offered line-by-line critiques of Corey's stories; Corey defended the stories with line-by-line rebuttals of Uzzell's comments. Finally, Uzzell wrote:

In my correspondence with you, Mr. Corey, I have done my very best to be of assistance to you as a literary agent, critic and friend. I engaged upon this without thought of pay for my services which I receive from others chiefly because Mrs. Stoll expressed the desire that I help you. I have read your writings also and believed and still believe that you have as yet undeveloped literary powers. I should have been very happy to have helped you arrive more quickly but I am afraid now you will have to go it alone. The first paragraph of this last letter, addressed to my defects as a critic rather than to my analysis and my suggestions, effectively closes the door to any further hope of success.

This does not mean any ill-will on my part or any reduced faith in your future; it just means that I can't help anyone who doesn't want to be helped. I have labored with you over your stories during the busiest weeks of my year, writing you at times when I should have been resting from day and night labor putting a little magazine and a book to press; I sincerely hope you have gained something from my efforts; as yet I myself have gained nothing.[33]

So parted Corey from this agent.

Later, when he sold "Their Forefathers Were Presidents" to *Story*, Corey wrote to Uzzell that because he had read the story, he would send the 10 percent commission as soon as he received his check, and he sent the $2.00 check in June.[34]

The Iowa Book

From 1934 into 1935 Corey corresponded with Simon and Schuster regarding his new Iowa book, but the editors declined it because, they said, of the pedestrian writing, humorless and unimaginative characters, lack of a central character, and absence of an absorbing plot.[35] The next year McGraw-Hill rejected it for lack of sufficient drama. Other rejections followed. William Morrow & Company rejected it for lack of intense characters. The next year brought more rejections. Corey revised the manuscript in light of each rejection.

Agent Elizabeth Nowell heard of Corey from critic and New School faculty member Gorham Munson and Dutch-born poet and novelist David DeJong, who was a frequent visitor at Paul and Ruth's country place.[36] She wrote:

> As I remember, you were the one at [W. W.] Norton's Caravan [an annual anthology that had published a Corey novella] party who said you were scared to death of agents—with which sentiment I honestly agree because I'm scared of most of them myself and would be even more if I were a writer. I flatter myself that my own way of working with people is not as cold blooded as some of the big agents, but on the other hand you may find that I butt in too much with criticism and all kinds of petty solicitudes. The reason I say this is because I have a suspicion from what I've heard that your book ought to be cut before it goes the rounds of the publishers. Of course it is impossible for me to tell until I see it. . . . It does seem as if you needed somebody to get you started toward the recognition you really deserve and

to get the book in perfect form and then start off with the big top-notch publishers who you evidently never have approached, instead of beginning with
the not so terribly good ones, such as Crowell and Morrow.[37]

Nowell wrote that she had read his work in the smaller magazines and
thought him a fine writer and that he could get much more money and recognition with a more careful buildup.

Paul responded that even though he had pulled himself up from the
obscurity of the little magazines to *Scribner's* and acquired a broad base of
favorable critical attention, "I'm at a dead-end now; my writing for the past
six months has been sterile. I can't seem to get my feet back on the ground
with my stories; approaching publishers with my novel gives me the jitters."
He continued that he liked to think that he could take criticism but admitted
that his difficulties with agents had been over that. He went on to explain the
kind of writing he wanted to perfect by using the *Scribner's* story, "There's
Always an Accident," as an example. Previous critics had told him to center the interest of the story on a single character. Had the critics accepted
his objective and pointed out ways to better accomplish it, he could have
followed their advice, but they had rejected his central goal. Ruth had promoted his novel at Harcourt, where she used to work. He agreed that the
novel should be cut by 50,000 words but hesitated to proceed without guid-
ance on what to cut:

> You see, it's a sort of fictionalized Middletown for the farm areas. That might
> not sound so exciting, but I gathered from talking with Munson that he felt the
> book to be alive and dramatic—in fact Munson seems to be the only person,
> with the exception of Jerre Mangione, who has read it and gotten the pattern
> and seems to understand the significance and bigness of the whole thing.[38]

He closed with a list of half a dozen publishers who had seen the 500-page
manuscript.

Nowell wrote to tell Paul that she did not think she could sell the story he
sent her, "Green Jackrabbit," because it lacked focus:

> It's sort of like a photograph of a crowd of people or of a group of buildings
> or some such taken from too close a range, so that all the figures are jammed
> in tightly and there's not enough perspective or enough focus to make a good
> composition (I mean not enough focus on one central point).[39]

She continued with a detailed point by point critique and indicated that she understood the role of the story in the larger novel. Corey responded:

> The matter of "focus" in the Green Jackrabbit story is confusing perhaps because I didn't make clear that the focal point is not a character, but the tractor. You see, the tractor is the focus around which the Middlewest must and will develop in the future and for that reason I tried to Pyramid the lives and problems of the group up to this machine.[40]

In 1935 the editor of the Peoria, Illinois, magazine *Direction: A Quarterly of New Literature* wrote to Paul Corey to accept his short story "The Farmer and the Gold Stone."[41] This was a story about how difficult it is to enlist people in a common effort unless they feel they have something personal to gain from it and then they act only for their own personal gain. Corey responded:

> Have just finished the second draft of a hundred and fifty thousand word novel, the first of a Midwestern trilogy, dealing with a section of Iowa farm country three miles square, showing the emotional, sociological and economic evolution of the area from a Marxian point of approach.[42]

The Magazine of Beverly Hills, California, accepted Corey's short story "Son of My Bone," which is about a man who accepts his wife, pregnant with the baby of a man she ran away with, when she returns. Corey wrote to the editor concerning his larger work:

> . . . . about the novel. . . . it's the beginning of a trilogy and . . . ends in 1917, which makes it somewhat untimely as to dates. Further-more it is more or less unique in its pattern . . . there are core chapters dealing with the central family in the farming section three miles square; then there are tied in with these more chapters, sketches dealing with other families in the square. That is roughly the pattern. I am trying to apply the maxim of "social evolution toward revolution" to this section.[43]

In a 1938 letter to Viking Press he explained to that *Three Miles Square* was not simply fiction; it was a work that no future historian of the Midwest could ignore, that he was trying to write an agrarian *Middletown*.[44]

At the advice of critic Granville Hicks, he sent a revised version to Harcourt, Brace. Another rejection. Corey changed his agent and continued submissions.

An editor at Random House, having heard of the work, asked to see *Three Miles Square*, and another agent wrote that they might want to represent it because writer Jerre Mangione had mentioned it. At Mangione's suggestion, the New York representative of Bobbs-Merrill sent the manuscript to the company headquarters in Indianapolis. Finally, Bobbs-Merrill accepted it toward the end of January of 1939. It had taken Corey four years and twenty-three publishing houses to find a place for the book.[45] In June Paul sent the revisions the publisher requested, and by the end of September, it was between covers.[46] That left two volumes to complete Corey's trilogy, and Bobbs-Merrill contracted with Corey for them.[47] *The Road Returns* came out in 1940 and *County Seat* in 1941. Meanwhile, Paul got a job with the Federal Writer's Project in Albany where his friend Jerre Mangione was an editor and had put in a good word for him.

The WPA Writers' Project

It was not an auspicious time to be a writer because the industrial economy of which writing was a part had collapsed just after Corey and Lechlitner began to get settled on their Hudson Valley farm and before Corey could become established as a writer. The election of Roosevelt and his New Deal provided relief not only for farmers but also for writers, among them, Paul Corey. Like the rest of the New Deal, the Writers' Project did not enjoy universal support in Washington.

Congressman Martin Dies, as chair of the newly formed Special Committee on Un-American Activities, was adamantly opposed to all of the Work Progress Administration's arts projects and, from 1939, attacked them all but focused on the theater and writers' projects. His accusations of communist activity and propaganda in the programs was front-page news and gained him the support of every opponent of the New Deal, including the Republicans and conservative Democrats in Congress.[48]

Employment with the Federal Writers' Project peaked in April 1936, with 6,686 personnel. By 1939, it was down to 3,500.[49] The project's major focus was the state guidebooks. Private publishers agreed to publish them, and there were 12,000 volunteer consultants, mostly college professors, who helped with checking the facts, editing the texts, and preparing manuscripts for publication.[50] Every manuscript was reviewed and rewritten at several levels, so all traces of individual authorship were erased by the time the guides got to press.

When the Massachusetts guidebook came out, a reporter noted that the Boston Tea Party had received nine lines; the Boston Massacre, five; and Sacco and Vanzetti, forty-one. That fueled a new press campaign against the Writers' Project and the New Deal, with headlines claiming that project personnel were all communists.[51] In fact the New York City project employed a number of radicals whose activities made the project vulnerable to such attacks. In August of 1938 Parnell Thomas of the Dies committee said the project was "one more link in the vast and unparalleled New Deal propaganda machine."[52]

Just to see whether his supervisor was watching, Corey once wrote a section of the New York guide in praise of Benedict Arnold. His supervisor was paying attention and excised it.[53]

Novelists and poets, in contrast to journalists, respected the past and could recognize and organize significant material and write it well. Corey was one of the project's best fieldworkers. Based in Albany, he had no difficulty getting Washington's approval for his copy.[54] His supervisor once exclaimed: "You're the only sonofabitch on the staff who can write the kind of shit they like in Washington."[55] About five years later, he would make good use of his rewriting skills. "For Corey writing guidebook copy was far more simple than working on a novel, but not as exciting, and he longed to return to his unfinished manuscript and to his farm," wrote Jerre Mangione, a coordinating editor, adding, "Corey was by nature a conscientious worker."[56]

Corey was critical of one of his colleagues in Albany for ignoring his work to write Republican campaign speeches. The colleague retaliated with a poison-pen campaign accusing Corey of being a communist. There was an FBI investigation, and Corey was cleared.[57]

Meanwhile, Ruth was working on the house:

Didn't get your wire till yesterday—but gosh, that's swell about the big novel, all right. Haven't got a letter . . . yet. . . . For heavens sake, I hope you can stave off revision etc. for some time. Don't let them rush you on it. I was so tickled when I heard the news I rushed right up to the house and planed a beam without thinking. All goes beaming there—but dam slow. I got to start working on the long ones today. Now that I'm getting limbered up some can work in the mornings too. The tools came yesterday; the scraper not much good with large handle on, so I detached the blade part and that works better.

All is o.k. in these parts so far, l except for this bitchy cold weather. It went down below last night; was still 1 below when I got up this morning, and I've got the old fire roaring away. Last evening the wind was terrible—just swept through the place. And it was about zero from 5 pm on.

. . . . How's Roosevelt? And what the hell do you have to do on the Tours? Prod 'em along so you can get back soon. . . . Hens are doing fine in spite of the cold. o—yesterday the pullets layed 22. . . . Certainly looks like that puttying on the windows will have to wait some time. Anyhow, I'm thriving on splinters and sawdust: . . . and I've swallowed so much wood lately I could sub for Charley McCarthy any time. Expect to shit a 2 × 4 soon.[58]

Just the next Saturday, she wrote:

As for the house, as long as we can move in come March, we should not worry about how much is finished. It would work out o.k. if that government job wasn't getting jammed up. . . . If we could just get someone up here end of next month that was any good at carpentering to help you work putting up walls etc.

Everything o.k. here. Too damn cold though. Me and the cat manage to keep warm—but guess I'll have to put anti-freeze in the chickens' water jugs. Dispensed with jugs entirely yesterday, and so have to dash over every once in a while and refill their pans. The last two mornings all the jugs and pans have been frozen up solid. But the hens are all right, and lay a good 2 doz eggs a day now. Have 6 doz ready this morning for customers.

. . . . Am still at the beams. Have just about finished doing the sides—and I sure hate to start on the underneath sides. Maybe with the scraper and sandpaper I can manage without breaking my neck muscles, but they sure are bitches. Nothing doing with the putty til it warms up. Have been feeling fine so far, except that my hands have chapped a bit.[59]

Corey wanted to get back to his novel, his wife, and his farm. Not wanting to lose someone whose work needed so little editing, Henry G. Alsberg, director of the Writers' Project, summoned Corey to Washington to ask him to reconsider. But after hearing Corey's story, Alsberg said, "Let the man go back to his wife and chickens."[60]

10

A COLLECTIVE STORY

They had no desire to erect artistic monuments as such. They had emerged from the farms, the village seminaries, newspaper desks, with a fierce desire to assert their freedom and to describe the life they knew, and they wrote with the brisk or careless competence . . . that was necessary to their exploration of the national scene.

Alfred Kazin, On Native Grounds, *207*

Editors Respond

Corey's correspondence with editors in the 1930s helps explain his concept of the "collective" story that he thought of as his contribution to naturalist writing in the tradition of Émile Zola. He described it as a sociological approach to the material rather than an individualistic one that would describe collective rather than personal struggles.

In February of 1935 *Partisan Review* editor Philip Rahv rejected Corey's story "Number Two Head-Saw" and suggested something shorter because as it was it would not fit with his program of publishing.[1] He expected discussions of proletarian literature from one of the several writers' congresses organized by the League of American Writers. Corey responded that he thought the story would fit Rahv's objectives:

> From the point of view of proletarian writing it seems to me that the "collective" story is pertinent to the principles of the Movement. The "collective"

DOI: 10.5876/9781646422081.c010

story attempts to show the dialectics of the class struggle rather than build up personal hatred as is the intention of the single character story which follows the old bourgeois pattern of situation presentation.[2]

He mentions Howard Rushmore, Frederick Maxham, and Josephine Herbst as examples of others who have worked with the same form and continues:

The single character story or perhaps I should say: the story dealing with a principle character is a very simple form because every detail is reduced to building a single individual and his problem. The bourgeois class being entirely individualistically principled has developed that pattern of story. The proletarian class stands for the group and for that reason I feel that the proletarian story must show the group, so I've chucked all "principle character" situations and all the stories I'm doing now deal with groups. But length seems to be an obstacle against my stuff being published and now I'm working on a condensation of group situations. I've approached no one of importance in the Party on this problem before, but if you've time I'd appreciate your opinion.

Marxism and the Party

Corey was well aware that in the 1930s there was a market for left-wing writing. This letter to Rahv, with its facile use of Marxist terminology, suggests that Corey's communism, unlike Alexander Saxton's, was instrumental rather than political. I think the evidence in chapter 8 is sufficient to indicate that he was not a member of the party. He didn't care where he got published as long as he got published. Corey was not alone. Literary historian Chester Eisenger wrote that writers did not read Marx carefully and even if they had wished to make his writings the theoretical basis for fiction writing, not much of his work was available to them in English. Thus when they did read him, it was in simplistic ways.[3]

The Communist Party organ *New Masses* rejected the headsaw story as too long and invited Corey to submit something shorter: "We liked this story a great deal and we'd like to print it. However, it's too long for us, it's really for a monthly. Why not try something shorter on us? We'd like to see more of your work."[4]

We Like It, But . . .

The editor of *Human Stories* wrote that his readers wanted stories of structured actions that would give them satisfying vicarious experiences:

> What you give them is complexity of vision, the complexity of your highly developed vision of life, a sensitivity they cannot follow with the same degree of thrill yours gives you. And thrill is what they want; but obviously different people thrill to different things; and with all the intense honesty of your relation to life, as I sense you, there seems to be between you, as writer, and the pulp public, a barrier of difference. . . .
>
> Take your "Number two Head-Saw" for example. By the time I got through with it my spirit went down. I could recover it by going to your implications; revolution and the steps leading to it. But the worker settling in his chair after a heavy day will find your fare heavy as his day, more so. And what he wants is to get his ease.
>
> I found your writing full of sensitive observation, snatches of vision I added to my store of experience. But I am afraid it was the writer in me that enjoyed these, a technician enjoying good honest work, rather than the editor who puts himself in the place of his readers.[5]

Another editor wrote that he had kept the submission so long because he liked it very much but, after discussion, had to reject it for being too long. He continued:

> Your point of view is exactly ours. We liked NUMBER TWO HEAD SAW immensely. On the other hand, we get quite a lot of heavy proletarian material. Have you any proletarian love story, or one which has beauty mixed in with realism? We would certainly like you for a contributor.[6]

Editing a semi-annual anthology called *New Writing*, John Trounstine praised Corey's writing but suggested the story could be cut by a quarter; he also advised reorienting it toward a central character and

> using the other characters as subsidiaries in the building up of the ultimate tragedy. At present one feels that one is getting a picture of three or four lives in vignette, but that the tragedy should be the major reflection in the facet devoted to Lerner [his candidate for central character] and the employment boss.

Though the theme is very Left, that could not preclude its inclusion in
NEW WRITING; but for American magazines, even of the Quality Group,
this may seem strong meat to some of the advertisers and readers.[7]

From the Marxist *Critics Group*, Corey learned in mid-1938 that

Number Two Head-Saw is a very solid story. We liked particularly the last
part of the story which concerns itself with the death of the man Lerner. We
are not sure that the little cameos in the first part of the story are so good.
The story is crowded. It reminds me somewhat of the version of Bushel of
Wheat, Bushel of Barley you submitted years ago to Partisan Review. Perhaps
Number Two Head-Saw should be taken and developed into a novelette like
the other story.[8]

Corey responded that he didn't understand the opposition to the story,
that he thought this was to be an anthology of new writing, and that he
believed a new approach to writing

has been and still is the portraying of groups of people confronted by a
situation and not single individuals confronted by a situation. I feel that this
approach is Marxian. Now, I'm not asking anyone else to completely agree
with me, but there is solid reasoning behind my conclusions.[9]

He wanted it published, he wrote, and if they still wanted it rewritten, he
could try to make "a Lerner story of it."

The editors responded that they were not opposed to portraying groups
of people confronted by a situation:

We do feel that for your space, you've crowded too many people in, the
reader finds it difficult to follow all. We are agreed that the best writing in
the head-saw story is found in the end. We don't want you to dismember
the story. If we can get the last episodes put into a finished form, I for would
one would push it among the other editors, and I do know definitely that the
other boys are waiting eagerly for it because, as I've written before, we are
very anxious to have something by you.

. . . . We all read your letter, considered your objections, and still do not
find ours weakened.

We don't want you to make a Lerner story out of any of your work.

We hope to speed up our work here and act on your manuscripts quickly.[10]

That didn't work out. Early in 1940 Conrad Wilkening, of the Wilkering and Son Agency, wrote that they liked the story and thought the *Saturday Evening Post* would like it because that magazine could use long stories. But the agent requested a different kind of change:

> Would you be willing to cut out the actual references to Kurtman and Sheig living with and having had children by Marcella? This part of their lives with her could be left to inference and not put in actual words and still be as strong. Any of the popular magazines taking the story would require this alteration.[11]

The agent offered to have his office prepare a new copy of the story if Corey would make the change. Apparently something went awry because later that year Corey has a letter from Lambert Davis of Harcourt, Brace accepting Corey's invitation for him and his wife to visit the farm and suggesting they would discuss "Number Two Head-Saw" then.

A Collective Story

Until now, "Number Two Head-Saw" has not been published because of the many objections about which the several editors wrote to Corey—lack of individual focus, lack of dramatic development, too many characters, and sheer length. Those reasons may be sufficient not to publish, but this story illustrates even better than his correspondence Corey's sense of what a "collective" story is. Furthermore, its setting is not the rural Midwest and its subject is not farming. In short, it gives some appreciation of Corey's reach. It also illustrates his sense of ethnographic description, which makes it attractive to anthropologists who are sensitive to the relationships among situation, behavior, class structure, emotional responses, and patterns of thought—culture.

The setting is much the same as that of Green Lumber, a sawmill in California where huge logs are floated in a pond to large head-saws that make the initial cuts in the logs to make them into planks. The workers live in barracks and eat in a company facility. Everything is ruled by the company, which tolerates not even talk of union organizing. Such an industrial setting is fraught with danger at every turn. The sawmill operates only when there are logs to process and is shut down when there are none. In the dialogue, Corey attempts to capture the accents of the immigrant workers from different countries. It was characteristic of Corey's naturalist writing to describe the setting, the work, and the workers in ethnographic detail.

NUMBER TWO HEAD-SAW

The prolonged blast of the six o'clock mill whistle shot upward with an expanding column of steam, expanding with the steam, to flatten out against the thick morning fog that hung over the plant and log pond, factory and drying yards. The imperious sound vibrated against the walls of the redwood houses and cottages, awakening the mill village. But this morning the shriek of the whistle meant more than usual: today Number Two Head-Saw was restarting after two years standing idle.

I.

August Gault turned over in bed, holding his arms out from the blankets as if to touch them would hinder the rolling of his thick torso. His wife, thin-flanked in her cotton nightdress, already had her feet to the floor.

"There's no coffee in the house," she complained. "Don't you go back to sleep." She paused at the foot of the white iron bedstead, vein-ridged hand resting on the gilt acanthus leaf decoration at the corner. "Two years of snuzin' late ain't good for any man."

Gault lay on his back and blinked at the ceiling. He was Number Two's sawyer, going back to his old job. He and his wife were down to their last slab of sowbelly; their savings in the redwood-pillared bank had been gone six months. Gault swung is feet to the floor and sat up as solid as a redwood stump. Just when they were about to starve or go on the County, zowie! Things start off again. What luck; He drew on the clean overalls his wife had laid out for him the night before and laced up his heavy mill shoes.

When he entered the small kitchen, his wife said again, "There ain't no coffee." She tucked the grey hairs beneath the rubber of her faded blue morning cap. Her pinched eyes watched his sullenly.

"We'll drink water." Gault washed the sleep from his meaty face. "Hanh!" he said, pulling his chair up to the table, "like old times, Becky." The blood of a workman marched through his body,—he had a job, he lived again. "The Company store's got to renew our credit now," he added.

"Well, if they don't and you don't put up a scrap for it, I will!" flashed his wife.

He hesitated a moment, then folded the slice of rancid bacon over and over beneath his fork and stuffed it into his mouth as if it were prime meat.

2.

The tramp of heavy feet resonated through the rambling corridors of "A" bunkhouse. Bill Kurtman sat up in his bunk, body brown and naked, pressing tough fingers against his swarthy cheek. He looked out of the window, trying to familiarize himself with his surroundings. The redwood walls of the administration building across the street loomed up darkly before his eyes; the fog obscured everything beyond.

He had just blown in the night before, having bummed up and down the Coast for months; sometimes hungry, sometimes fed. Periodically, since the California Lumber Company laid him off two years ago, he had returned; but Pendegrass, the employment manager, sent him away each time with: "No, nothing doing, Bill." Last evening he was offered his old job,—"setter" on Number Two Head-Saw. Well, they couldn't get good setters every day.

Kurtman flung back the heavy wool blankets and combed his stiff hair with his fingers. And they weren't willing to pay a man what he was worth to 'em either! His ears stood out large and red from the stubble of his close-cropped head. He'd had a good supper at the mess hall last night; he'd had a good bed for a change and there were prospects of a good breakfast ahead of him. The Company fed its workers well; that much could be said for it, although they docked a plenty from the pay-check for board. His body flushed with a certain contentment as his blood seemed to reach back into the past, grasping the broken stands of old habit, fusing them again into familiar routine. He pulled on his belted dungarees and without lacing his shoes, flung a towel over his arm and strode down stairs to the bath house.

With a slant-wise glance he appraised a short fellow bathing in the basin beside him. "I'm new here," Kurtman said, "What's the gab around the mill?"

"You'll earn your money," replied the fellow, cheek patched with soapsuds half-turned toward Kurtman, eyes searching, pupils dilated like a cat's when frightened. He added: "And damned little money!" then plunged his face into cupped hands full of water smelling of redwood.

Certain things that had happened to Kurtman since he last worked for the Company rankled deep in him like sand grains beneath the glide of his muscles. Hungry men herded, driven, beaten; the roar of riot guns; the bite of tear gas: he had learned about such things. Suddenly his eyes brittled with

brown fire. A recent happening leaped into his mind. He'd smashed the jaw
of a dock-owner's hired thug. Then came flight—the darkness of alleys and
his memory of the feeling of raw board fences beneath his touch. "Hah!" was
the exclamation his thoughts formed. "You said it, brother," he growled to
the man beside him. "Damn little money they give us!"

He swung out of the bath house, clumping up the stairs, heels of his
loose shoes banging, he put away his towel and got his shirt. The muscles
of his thick fore-arms danced and rippled as his arms swung backwards and
forwards. He heard the mess hall bell clang and quickened his steps.

3.

Half a mile up the river from the mill village and across it, the columns of
the standing redwoods seemed to hold up the fog, their high branches hidden.
About the base of the trunks lay a mat of dried six-foot fern fronds and foli-
age. The foot-thick bark of the trees streamed with moisture, and fat yellow
slugs, as long as a middle finger, crawled fluidly over their ridges and canyons.

Near the foot of a great tree was a grave-shaped mound of fern fronds and
needles. After the blast of the mill whistle, the mound quivered a little, then
was still; then quivered again. At last a bony hand pushed away the heap and a
pale lean face, covered with a dark patchy beard—the paleness of the flesh and
darkness of the beard each accentuating the other—appeared like a resurrec-
tion. The brown fronds fell away from the angular shoulders supporting a faded
blue serge coat. The man shivered and hugged his arms V-wise across his chest.

Then he stood up amid the great redwood trunks and head-tall dark green
ferns like the last member of a species nearing extinction in these primeval
depths. Again a tremor jerked his body. He looked with blank eyes at the
nightmarish bark of the trees. He seemed to be listening again, searching
for the echo of the mill whistle as if that would have some meaning for him.
Then with faltering steps, he passed through the openings in the frond-walls
of the undergrowth, letting his animal senses of direction guide him out of
this Karnak of redwood columns.

At the river bank he stopped and washed his hands and face, and the cold
water against his corpse-white skin revived the sluggish blood and warmed
him a little. His shivering let up for a time but recurred occasionally with nervous insistence. He crossed the foot-bridge over the river and walked up past
the Redwood dance pavilion to the pike. For a moment he hesitated before
stepping out on the paved road, and considered going around to the back of

the pavilion to see if any scraps of food might be found in the refuse. But his mind seemed unable to make a decision and his legs carried him unsteadily on toward the mill village.

4.

The workmen left the mess hall in twos and threes; some walked with quick steps, others slow and reluctant, but all flowed in the eddying stream down the pike toward the mill. Emil Sheig, pondman for the head-saws, heard talk behind him: "Hello, Bill!" "Bill Kurtman! By God!" "Didn't see you sneak in to the feedbag!" "How're yuh, Bill?" Sheig stiffened, his red neck-hair bristling. "Dat punk back!" he muttered between his teeth.

His hob-nailed pond-shoes crunching the cinders, Sheig turned from the stream of workers and set out on the shortest route to the pond-saw. He was one of the morning shift on the pond. Other pondmen sauntered just ahead of him but he didn't hurry to catch up with them,—nor did he look back over his shoulder toward the flow of millmen. Kurtman back! Dat punk! Why hadn't he died mit starvation? Ach! A scowl bronzed with shadows his red-stubbled Latvian face. The return of Kurtman, the setter on Number Two Head-Saw, revived memories of a situation two years old in the pondman's brain—a situation that he had hoped was over. Head bowed, Sheig walked with the nimble steps of a man used to treacherous footing; his thick hands stained black with the redwood swung clenched at his strong thighs.

Would Kurtman go again to Marcella Condetta? Sheig rubbed the ridge of his broken nose with a quick reminiscent gesture. He'd fought Kurtman to a draw over Marcella. Ach! They had mashed each other up all right. But Marcella was his woman! Didn't he give her money when he had it? When her folks couldn't get relief from the County?

Sheig crossed the bridge over the canal that joined the ponds of the upper and lower mills and turned down the footpath toward the "B" mill pond-saw. Just last night Marcella had been making talk about marriage. He had laughed—partly to tease her, partly to push aside this threat to his freedom. He didn't know that Bill Kurtman was back then. Now he brooded, maybe it was better he shudt marry wit her.

One of the three pond-men busy at raising and anchoring "sinkers," the heavy butt-logs which would not float, shouted to him: "Hey, Emil, see your friend Kurtman's back."

Sheig did not answer.

The pond-sawyer shouted: "Snap out of ut! For Chrissakes! Number Two's startin'! We're short-handed anyhow, without you takin' your own goddamn time getting here."

Sheig said nothing. He took his pick from the wall of the pond-house and stepped nimbly upon the backs of the floating logs.

5.

Hooking his thick thumbs under the bib of his overalls and clasping his fingers over his empty watch-pocket, Number Two's sawyer, August Gault, set out down the gravel path from his cottage to the pike. He reached the pavement just as the man from the grave of fern fronds came along, going toward the village. One glance into the bum's cadaverous face gave the sawyer's stomach an unsupported feeling. Looked as if that guy was about to pull the dirt in on top of him.

Gault joined the straggling flow of workmen on the other side of the highway, nodding to those he knew, and walked on toward the mill, forgetting quickly the starved features of the bum. He hid the memory of this walking head-line of hunger and despair by thinking with all his will how lucky he was to be going back to work. But his body seemed to refuse to forget: he belched suddenly the disagreeable flavor of rancid bacon.

The fog had lifted a little and Gault could see over the wide log pond,— acres of logs. He saw three animated specks working on the dark surface: the ponders [workers managing the logs floating in the large pond] rising sinkers; he noticed a man walking slowly along the distant bank of the pond: that was the Head-saw's pondman, Emil Sheig. Galt's eyes sought with eager sureness the dark opening in the mills, where the log slip went up out of the water to the head-saws. Grey columns of smoke pushed up the ceiling of fog, merging with it, above the stacks of the power house. At the brow of the hill on the highway, he looked over the saw-toothed roof of the factory and sheds stretching away to the wall of young redwoods in the distance. Just like old times! He thought. Then he wondered who employment manager Pendegrass would get him for a setter. He didn't know that Bill Kurtman was back, but Kurtman was the man he thought of when his mind turned to "setters"—Bill Kurtman was a great "setter," and a great guy. Gault hurried his steps toward the yard entrance of the mill.

6.

The bum from the grave of fern fronds didn't pass unnoticed by the stream of workers on the pike. A young mono-rail operator started deviling him across the pavement: "Hey, fella, when'd you eat last?"

Others among the younger workers took up the fun: "The buzzards're lookin' for you, guy!"

The wave of vicious banter broke around Number Two's setter, Bill Kurtman, as he strode along. Without slacking his pace he turned his head and his dark eyes took in the man's appearance.

"Shut up, you bastards!" he shouted and the jeering stopped. "You stiffs ain't so goddamned lucky. If you weren't makin' plenty of jack of the Company you'd be in the same fix."

The bum lifted his head and looked toward the line of workmen. He smiled, but the expression was like a ghastly reflex of muscles on a dead face. The saliva suddenly flowed in Kurtman's mouth, thin and watery, as if preparing something nasty to be ejected. He spat and strode on. He told himself that he wanted to be on the log-deck before the starting whistle blew. Pendegrass, the employment manager, had told him that good old Gus Gault would be his sawyer on Number Two.

7.

Reaching the village, the bum glanced up and read the sign in gold lettering "National Bank" above the entrance to the oblong building, columned on four sides with redwood pillars, which stood at the jog in the highway. It seemed as if he were sniffing the air rather than reading, for he turned across the street abruptly and shambled over to the kitchen door of the mess hall. The cook stood behind the screen, leaning against the frame.

"What's chances for a bite to eat?" the man inquired.

The cook took his fat arm from its angular bracing position against the opposite frame of the door. "Sorry, brother, it's against the rules of the Company."

"I'll work for it," continued the bum. "Wash dishes" Then he stammered: "Just—just up from the Valley—ain't—had nothin' to eat for a couple of days."

There was a rumbling sound in the cook's throat and the bum drew back as from a dog growling. The cook said: "You sure don't look well fed, brother, but it's as much as my job's worth to give you a crumb."

The bum turned away.

"They've been takin' on some help over there," the cook added, gesturing across the street toward the administration building. "You might try 'em. Pendegrass—employment manager—'ll be down in about an hour."

The bum looked blankly at the crowd of men already waiting beneath the sign: "Employment Office."

"Wish I could give you a snack," said the cook.

The bum mumbled a "thank-you" and shuffled over to the administration building.

8.

The seven-thirty whistle screeched and Number One Head-saw caught up and carried on the shrill sound as it whirled into action. Number Two's carriage was empty and there were no logs on the deck.

Kurtman and Gault hugged each other and slapped each other on the back like brothers meeting again.

"How're yuh, Bill?" "How're yuh, Gus?" they shouted, forgetting that they couldn't be heard above the scream of Number One, forgetting that they could read each other's lips as they did in the old days.

Then the setter stepped back and surveyed the sawyer critically. "You ain't got thin, Gus," he said.

Gault laughed and his body, paunch-soft with idleness, quivered. "Not exactly," he shouted, "but if things had gone on much longer the way they were, my old lady and me ud uve been barked to the bone. We're lucky. Damned lucky for me to get back on the job before we had to go on the County."

"Lucky!" snorted Kurtman, the lines of his face setting hard. "I don't call that luck! Lay you off till all your money's gone; then put back to work so's you can just pay your debts. By God, a man's a right to earn a living all the time."

"Better watch out, Bill," cautioned Gault. "Ain't safe to talk like that around the mill."

He didn't shout now; both of them suddenly seemed to realize that shouting was unnecessary and their eyes were fixed on each others lips. Then Gault turned away to test out his saw and the pedals which operated the steam "niggers,"—enormous steel arms which handled the great logs like match sticks on the carriage. He was disturbed by the fighting look on Bill's face. It reminded him of the look in the eyes of his wife, Becky, this morning when she talked about renewing credit at the Company store. He shook his head: he wanted to—it was best just to feel lucky to have one's job back.

Kurtman walked out on the log-deck between the two head-saws. Memory of the bum on the highway flashed into his mind, bringing a taste of iron to his tongue. Gault was right, he thought, a man had better keep his mouth shut. Above him the mill structure rose dark and shadowy, its heavy cross-beams obscured in spite of the row of windows in the upper end of the building. The log slip slanted down from the deck into the pond, the prongs of the jack-ladder rising sharply in the slot of the log slide. The end of the mill, below the eaves, was open to the pond and the thirty-five acres of logs stretched away to the rows of small bungalows which stood at the edge of the water on the opposite side. Kurtman watched the carriage of Number One start forward and waited tensely for the shrill voice of the saw as it bit into the solid wood.

While Gault tested the waste-wood conveyor and the rollers which carried the plank to the re-saws, he defended the company hotly to himself. He had to do it to keep from facing the fact that he had been on the brink of hunger and that the only reason he was able to keep a roof over his head, with the mortgage interest unpaid on his cottage for eighteen months, was that the small house had no value to the Company, who owned it, as long as times were bad. A frown puckered the sawyer's heavy features. Who was he to criticize the Company? They owned the plant and the land and the timber, didn't they? They had a right to do what they pleased, didn't they? And he had his job back. Now he could start paying up the back interest and save a little for the next rainy day. Things had just worked right for him. He wasn't going to kick. No sir!

The carriage of Number One shot back for the fourth time. Kurtman turned away. Where was the pond gang? Why the devil weren't they along to get a log for Number Two? He jumped to his position on the carriage and swung the lever on the setter's dial, testing the working of the mechanism.

The mill foreman appeared, checked in the new crew; then hurried away to find out what was delaying the pond gang. Gault came up beside Kurtman on the deck and the latter complained at him: "Damn this loafing! If I had a pick, I'd pole a log over here myself."

9.

In front of the administration building, the crowd of unemployed had increased. Hulking Pendegrass came swing-striding up the steps, pushing brusquely among the throng. A hush cut through the buzz of talk at his appearance. He stopped suddenly, hand on the door latch, and spoke to a sallow-faced man standing hunch-shouldered with the others: "No use your coming here, Blake. The Company won't take you back. You talked 'organization' once too often to the men." His voice was stern but his gaze unsteady.

The man lifted his head, mumbling: "Won't say a word again, Mr. Pendegrass, ah promise. Mah wife and kids—"

"Should have thought of them before," snapped the employment manager. "The Company won't stand for trouble makers."

"But, ah promise—"

"Get off the Company's premises!" thundered Pendegrass, his voice quavering. The man moved out of the crowd without a backward glance, shoulders squared. He didn't come licking the Company's boots because he was beaten,—but his wife and kids . . . Pendegrass swung into the building and the crowd surged toward the door after him. The bum, having joined the waiting job seekers after his talk with the cook at the mess hall door, was well down the line, staring listlessly at his feet with a dead look in his eyes. The man, Blake, sauntering away, brushed his lean arm, but neither noticed the other and the forward movement of the crowd toward the office jostled the bum with it.

The employment manager sat down at his desk and thumbed over the mail idly. Let 'em wait. Still ruffled by his recent encounter outside, he stared blankly at the panorama picture of the California Lumber Company on the wall. A slight twitching high up on his cheek pulled back the corner of his eye to a pained expression.

An hour earlier, he had stood for five minutes by his wife's bedroom door, listening. A feeling of crushed loneliness and hurt pride bore him down. If he could just hear her move, perhaps cough a little—but there was not a sound. He saw little of Beth these days. She never got up to breakfast with him; she was away much of the time; her one ambition seemed to be to win the mill-village flower show and impress Mrs. Logan, the Company president's wife.

As he sat here at his desk now, he recalled the cold, hostile expression he usually read on her face. She always seemed to look at him as if he were repugnant to her; she resented his being only an employment manager in the Company. Why didn't he get on? Why wasn't he somebody?

The sound of the jostling crowd attracted his attention. Let 'em wait. His secretary came in from the adjacent office.

"Collins, on the pond, called in and said he needed more help," she informed him.

"More help, nothing," Pendegrass growled. "Let him get more work out of what he's got."

The secretary left him. Again he thumbed the mail. At last week's executive conference, Logan and Schroeder were hot about taking on more help—starting Number Two was all right but they felt that auxiliary help was unnecessary. Make the help they had carry the additional load. Caxwell, the general manager[,] had objected. Pendegrass took a short quick breath: Caxwell was strong in the Company but Logan was president, Schroeder first vice-president. Again Pendegrass saw himself standing in front of his wife's bedroom door; the memory of the pale green glass knob was suddenly vivid to him.

Then one by one he let in the unemployed. He questioned them curtly and hustled them out again. There was no work. The bum reached the hallway of the building finally—he was next up. His remaining strength surged in him, combatting a weight of hopelessness.

The telephone jangled in the office and he heard the employment manager arguing with someone:

"A man on the pond? Collins called before I came in. No. I'm not going to send one. Well, what the hell do you want a man there for? Get more work out of those punks! Yes, I know both saws are working,—I know that! That don't matter,—get more work out of 'em! Say, you tryin' to tell me my job." There was a pause; then: "Well—ah—well, don't get hot about it. Won't be necessary to take the matter up with Mr. Caxwell."

The receiver of the telephone clicked down on the support.

10.

"Here comes a pickman," shouted Kurtman. "By God, it's about time."

He ran down the side of the slit to the pond; then stopped. He recognized Sheig. Gault remained on the deck, thumbs hooked under the bib of his overalls. The setter and the pondman approached each other like strange dogs.

Kurtman growled: "You fellows playin' mumbly-peg behind the pond-saw?"

"Zhort handed!" snapped Sheig. He stepped out on the floating logs and began poling a huge timber. It moved slowly, like a giant hippo, toward the slip. Sheig's teeth were jammed shut: for vy didn't he lam dat punk mit his pick? Kurtman grasped a sliver of bark and grudgingly helped guide the log into the throat of the jack-ladder. He consoled himself that he'd done a fine job, two years ago, of arching that hunky's beak.

Gault on the deck started the power on the jack-chain. The heavy links clanged over the guides. A hook caught in the log but slipped its purchase; another dug into the wood and dragged the trunk a little way out of the water. Sheig and Kurtman, one on either side, steadied it. A third prong caught, and the tons of wood, barked, curved sides, glistening black and water-soaked, began to climb the slip. The mass rose shoulder high between pondman and setter and they scowled at each other over this shiny barrier. The log reached the top of the ladder and a skid automatically rolled it onto the deck.

The sawyer stamped on the pedal operating the "hill nigger": the great arm shot out, gripped the log and rolled it thundering upon the carriage frame. The "dogs" were set quickly. Kurtman stood by the setter's dial, the knob cool and familiar against his palm. He no longer thought of Sheig or Sheig's woman; his attention was fixed to the log on the carriage. Gault threw the switch to the saw and the steel band began to move, its sharp teeth becoming a shimmering ribbon of blue mist. He glanced at the setter. A thought flashed through his mind: Bill didn't think he was so lucky to have his job back; Bill didn't think the Company was doing him any favor. Damned poor wages!

He signaled Kurtman and the carriage advanced on the saw.

II.

Pendegrass appeared in the door of the employment office. "Who's next?"

The bum rose unsteadily, grasping the redwood arm of the rough redwood bench for support. The employment manager looked him up and down quickly; then beyond him to the others pushing at the entrance of the building.

"Come on," he ordered gruffly.

Inside the office, Pendegrass sat down in his swivel chair. He folded his arms and looked at the emaciated figure before him, a half smile, half sneer framing his heavy lips. "Name?"

"Ralph Ler—Lerner."

"Ever work on the pond?"

The young man's glance refuted the sneer on the employment manager's face; then rose to the panorama of the California Lumber Company. He had never touched a pond-man's pick.

"Yes," he grunted.'

Pendegrass sniffed contemptuously. "All right, report down at the pond-saw. They need a man there."

To Lerner the panorama of the Lumber Company suddenly reeled before his eyes. Waves of faintness radiated outward from the void of his stomach. He felt as if he were going to vomit. But one rigid point remained in his consciousness: he had a job.

"Thank you," he said. From that solid spot another pushed outward against his weakness. "I—I haven't had my breakfast yet. Would it be possible—"

Again the slight twitching speared high up on Pendegrass' cheek. The vision of himself standing forlornly, imposingly before his wife's bedroom door flashed into his mind.

He snapped: "Breakfast's been over two hours. They're in a hurry for a man on the pond." His fingers fumbled for a cigarette. Lerner still hesitated. "If you don't want the job—" the manager continued meaningly. He lighted the cigarette and laid the burned match in the green agate ash-tray on the desk.

The young man turned unsteadily toward the door. Then Pendegrass relented a little: "I'll have everything fixed for your dinner." His hand gestured fatuously. "Stop in this noon, the girl'll give you your meal check and room number."

"Thank you."

Lerner stumbled out of the administration building, vaguely feeling that there was something to look forward to: dinner. He started toward the canal bridge and the board-walk to the pond-saw. What did it mean working on the pond? He thought no farther, unequal to the effort. The sunlight splashed through the rolling clouds of fog and warmed the outside of his body.

12.

At Number Two Head-Saw, the thin line of steel mist struck the log with a whine; then rapidly swelled to a shriek. Gault's eyes centered on the advancing cut, riveted to the fierce movement and the steam-like spraying of waste-wood into the refuse conveyor. Lucky or not lucky; poor pay or good—who was he to complain? Vexing problems were drowned in the screaming of the saw and his body vibrated with a feeling of power,—he was a sawyer; he knew his job; he was doing his job.

Kurgman rode the moving carriage, his head turning as the cut approached and passed him. His ears tingled to the shrieking which numbed all his nervous excitement. The cut completed, the curved cant banged upon the re-saw rollers; the carriage crashed back against the hydraulic bumpers ready for a new cut. The sawyer signaled for a three-inch flitch; Kurtman swung the lever on the dial. The mechanism shoved forward the log and again the carriage advanced on the saw.

In the brief lull, Kurtman flashed a glance over the breast high timber in front of him and around the mill. The Company had the workers by the short hairs, by God! The rising complaint of the saw cut short his thoughts. The plank banged to the conveyor. The carriage shot back to position leaving a clean reddish surface to the down-thrust glow of the flood-lights overhead. He took Gault's signal for the cut, swung the lever on the dial; and once more the carriage moved forward.

13.

The board walk skirted the upper end of the pond to the railroad tracks where the logs were brought up from the timber. The roadbed near the water was on a slant so the logs would roll from the flat cars of their own weight. The bum, Lerner, followed the slanting cinder path over the sleepers, his feet nagged by the raised, narrow-spaced ties. He could see farther ahead where the board walk started again, this time directly across the pond, and some distance along it the small house of the pond-saw. The tenor screaming of the steel, cutting the logs into lengths for the mill, reached his ears. The water was open in the upper half of the pond; a dark greenish blue, colored by the vitriol put in to keep down the scum and fungus—a bellyful of that water would make a person plenty sick. Stretching far ahead, Lerner saw the curved backs of hundreds of logs, like a strip of broken corduroy, as they crowded toward the mill end. The three men raising sinkers watched him go past but said nothing.

The board walk, built on short piles, curved around a great heap of logs on the right—the reserve log pile—which hid the mill from the pond-saw. The dark water lapped the boards, dark-patching the wood, and the movement made Lerner dizzy. As he neared the pond-saw a log held against the wharf suddenly broke in the middle and the two ends bobbed to the surface. The scream of the saw subsided in a fluid grunt. A pondman poled the cut logs clear of the blade and into a stretch of open water, walking agilely on the backs of other floating logs. Lerner watched him, his breath coming quickly.

The sawyer's grey-whiskered face appeared in the square rear window of the pond-house. "New pondman?" he asked curtly. Lerner nodded and submitted to his cursory but critical glance. Sheig approached along the walk from the mill, pick over his shoulder, still grumbling inwardly against Bill Kurtman.

The sawyer asked: "Name?"

"Lerner."

"Been on a pond before?"

"Ye—yes." Lerner was fascinated by the man on the pond who calmly poled the halved logs toward the mill jam.

"You don't look it," growled the sawyer; then: "Here, Sheig, 's a helper for you."

The pondman glanced quickly at Lerner, first at his face; then at his feet. A look of contempt darkened his flat Latvian features. "Thereza pick." He gestured toward the row of slender shafts leaning against the whitewashed wall of the pond-house. Lerner took one nervously and held it like a lance. It seemed enormously heavy. Sheig turned back toward the mill: what did this punk know about a pond? Ach!

Lerner followed him dumbly, the uneven surface of the log-backs on either side of the walk confusing his vision. The point of his pick caught and almost threw him among the floating timbers. His muscles, hunger-slack, were slow in righting the off-balance of his weak body.

Sheig looked around. "Where'd yuh ride da logs?"

Lerner's consciousness jerked; he found himself mumbling: "I don't know"; then he said: "Up the—the Coast."

"Hah!" the pondman stopped. "When you eat before, brooder?"

With an effort, Lerner pulled himself together, trying to parry Sheig's keen glance. "Two days ago—guess—" He broke off.

"You no have breakfast?"

"Mr. Pendegrass said breakfast was over." The explanation came easily.

"Sohn-uf-beesch! Cook can make breakfast all time!" The pondman shook his pick with a gesture of exasperation. "It is a rule. A man can always eat before he go to work."

For some reason, Lerner felt that he should defend Pendegrass. "He told me he'd have everything fixed so's I can get dinner." It pained him to talk about eating, otherwise he could forget his hunger.

"Hah, damn right! Sure you get your dinner, hah!" Sheig ripped out a string of Latvian profanity, lashing Pendegrass, and walked on. Lerner followed him, head hanging as if he had been scolded.

14.

Number Two Head-Saw had made four cuts from the side of the log. Gault stepped on the "hill nigger" pedal, the huge arm rolled the heavy log on the carriage like a pencil between the fingers. The weight shook the frame with hollow thundering. Kurtman adjusted the timber for another cant cut. In spite of concentration of the work both he and Gault were aware that Sheig was bringing a second log to the deck and that he had a helper. It was Lerner's awkwardness and frantic energy that attracted their attention. He was trying to show that he was worth his salt. No one was going to call him a loafer. He pushed with his pick against the log as the ladder pulled it to the deck; then threw all his weight behind it, trying to roll it from the chain. The iron skid tipped it easily toward Number One's carriage: Lerner stumbled, almost fell, and looked about foolishly.

Gault threw the switch, restarting Number Two. He recognized Lerner as the bum he'd met on the highway. That fellow was in luck too! He'd looked pretty near down and out on the road. This consideration of the bum pushed aside a little train of thought he had been playing with. Supposing, he had said to himself, just supposing the Company store refused to renew his credit until he'd earned his first pay check. He knew there was no chance of that. But he found a little amusement in worrying himself with the idea. Just supposing they did. He bucked himself up by thinking: well, Becky and he'd put up a scarp the store would remember. He flashed a sudden defiant look over the log-deck and saw the bum working with Sheig. The fact that the bum had a job seemed to add to his assurance that he'd have no trouble getting credit. He signaled for the cut.

Kurtman replied "Okey." A frown spread outward from his dark eyes. What was that bum doing in the pond-crew? He didn't look like a ponder—anybody could see that. Kurtman felt the carriage tremble beneath him. He cast a quick glance at Lerner's thin shoulders; then saw Sheig from the corner of his eye. He felt a sudden fury: they wait till a man's almost starved to death, then put him at a tough job like the pond. The blade of the saw struck the log and absorbed his fury; he wondered vaguely if Sheig was still stuck on Marcella Condetta. The carriage continued to move and Kurtman's eyes centered on the blue ribbon of mist screaming into the log.

Sheig returned to the pond: Lerner tagging after him. They had to get another log to the deck for Number Two. The Latvian, riding the jam, started a log out into the open water in front of the slip. It nosed into the end of a trunk and from his location he could not pole it loose. Lerner tried desperately to shove it free but he was unable to reach it. He trotted back and forth frantically, bending over, jabbing at it with his pick; then he stepped gingerly to the logs and felt them settle beneath his weight. But they seemed solid. His courage grew and he approached the jam. He started prodding at the obtuse log.

"Push um back!" Sheig shouted.

Lerner threw all his strength against the pick. The log moved back and free. It drifted a little to one side. Suddenly the timber beneath his feet began to turn. Instinctively, he stepped higher on it, which only increased the movement. The log next to it began rolling. The wet sides came over and the young man's work-soled shoes slipped on the slimy wood. He dropped the pick, struggling fiercely to keep atop the slowly but persistently revolving logs which continued to gather momentum. He lost his balance and fell on his back between them. His hands clutched at the slippery sides. The log on the open water side gave way a little and his legs went through into the cold pond. He screamed. His eyes stared wildly at the sun breaking through the fog overhead. The great timbers went on rolling, thrusting him downward into the water like a bundle of rags through a clothes wringer.

Sheig came leaping toward him across the logs, skirting the open water in front of the slip. Lerner's bony fingers clawed the slimy wood. His shoulders were forced under. The logs kept turning relentlessly. The huge butt he had freed crowded in by the current against the logs he was between. His head was being rolled down. Sheig yelled something in Latvian. Lerner's wildly waving hands knocked aside the pondman's pick thrust at him to cling to.

.

The logs closed in and slowly rolled the twitching, wriggling fingers down into the dark water.

Kurtman, bracing himself for the backward lunge of the carriage, looked toward the end of the mill and his eyes, seeing beyond, glimpsed those hands disappearing, slowly, yet with awful speed between the rolling logs. He jerked the emergency stop to the carriage and vaulted the breast-high timber to the log-deck. Gault's eyes flashed him a look of crazy surprise. And hands fumbling, shut off the power.

Sheig was poling with all his might at the jam. "Unner there," he bawled at Kurtman.

The setter leaped out upon the logs, grasped Lerner's pick and began working like a demon. The sullen look that had clouded Sheig's face all morning was gone. At the moment, Kurtman was any man to him. The faces of both men were sharp cut with a grimness of fighting fury as they struggled to open a space between the tons of water-soaked wood and save Lerner. Gault came lumbering down the slip side. Number One stopped abruptly and the crew joined the others at the pond. The foreman came stuttering up: "Wha—wha—what's the matter?"

Number One setter said: "Man under the logs."

Gault said: "Ought to call the ambulance."

More workmen, coming from farther back in the mill, crowded along the bank. Sheig and Kurtman broke the logs apart but their footing growing unsteady, they had to pause to balance themselves. In the narrow strip of water a sodden serge shoulder appeared.

"There he is! Grab um!" came shouts from the crowd.

The timbers came slowly together; the two men thrust them back. The millmen thronging the pond edge were tense, leaning forward like a human hedge beneath a wind. The foreman had gone to call the ambulance and Pendegrass.

Kurtman, riding dangerously one of the logs that had moved over Lerner's body, raked his pick up and down in the water. His hook caught in something soft. "I've got 'im!" he yelled to Sheig, who was keeping the other log back. Then he began to draw in his pick.

15.

The whole plant seemed abruptly quiet. The re-saws had stopped. The sorting table web no longer moved; the crowd at the pond increased, jostling and shoving, extending up a slant of the log-slip. Everyone craned and stretched to see better. Again the serge shoulder appeared in the water. Kurtman, with a quick twist, unhooked his pick, and kneeling, grasped the soaked coat collar. With a jerk, he hauled the limp body up the back of the adjacent timber.

Lerner's eyes were closed and his face had a skim-milk bluish cast as if tinted by the vitriol water. A dark bruise showed on the side of his head. Sheig lifted the body toward the walk. Men stepped out on the precarious floor and carried it on, laying it carefully on the boards. The crowed edged closer.

"He's a goner," somebody said.

"Don't look mor'n a skeleton," remarked another.

A sorting table man said: "If he ain't drowned, that vitriol water'll make him sick aplenty."

Kurtman and Sheig turned the body over. The setter grasped Lerner by the hips, raised him like a bag of meal and shook him. Water gushed out of his nose and mouth. The arms flopped limply against the boards. Laying the body again, Kurtman worked the chest and arms in the slow movements of artificial respiration. The bony shoulders and ribs made him feel as if he were working with some wooden contraption.

"Don't look like much of a ponder," observed Gault.

Sheig, still panting, growled hoarsely: "Neffer see a pond—kid—come mitout his breakfast. Hain't et for two-three days."

"What's that?" snapped Kurtman. "Didn't Pendegrass let him have his breakfast?"

Sheig grunted: "No."

The setter's large hands moved the match-stick arms with clumsy rhythm. "Good God!" he said.

Sheig's remarks passed back through the crowd.

"Kid was starved?"

"Pendegrass wouldn't give him his breakfast."

"Hadn't had anything to eat for a week."

"By God, how long's it been since the Company's so poor they can't feed a man before sending him to work?"

The resentment and indignation swelled and spread among the men; faces here and there stood out suddenly, eyes angrily threatening. The dull distant roar of the mill seemed to echo the mounting anger.

16.

The employment manager crossed his log-deck frog-honking: "Gangway! Gangway for the stretcher!" The workmen opened a passage quickly, watching with sullen faces. "What's wrong? What's the matter?" Pendegrass kept repeating as he approached the men around Lerner. A young doctor, carrying stretcher and pulmotor, followed close at his heels.

Kurtman stepped back. The doctor took his pulse, made a hasty examination, shook his head and attached the pulmotor.

Pendegrass said: "How'd it happen, Sheig?"

"He fall in!" The pondman's face knotted with fury. "You not let um have breakfast. He did not eat for—long time."

"How'd I know he was hungry?" retorted Pendegrass.

"Take a look at 'im," drawled Kkurtman. "You don't need brains to tell he was hungry."

"He ask for his breakfast," accused Sheig. "He tell me."

The employment manager's face became a swollen brick red; the pores stood out distinctly like pin punctures. He kept telling himself not to argue with these stiffs.

"You killed um!" yelled Sheig like a madman.

The grumbling grew in the crowd.

"Pendegrass wouldn't give the kid his breakfast."

"The kid was starvin'."

"Somethin' ought to be done about it, by God!"

The employment manager stepped hastily to the opposite end of the prone figure, placing the body between himself and the crowd.

Gault's voice boomed: "What's the meaning of this Pendegrass? The Company allows everybody to have breakfast before work. We won't stand for this kind of treatment."

A graduating bead-work of sweat covered Pendegrass' brow. He was cornered; he should have fed the kid. "An—unfortunate mistake, Gault," he stammered.

Kurtman, eyes blazing, swung upon the millmen lining the log slip and revetment of the pond,—tense faces, restless arms and crunching feet. "You fellows!" he roared. "How long 're we goin' to let bastards like Pendegrass get away with givin' us the dirty end of the stick?"

"Throw 'im in the pond," voices answered, "See how he likes it."

"Men! Men!" Pendegrass' tone rose stridently. "Go back to work—"

"Shut up!" ordered Kurtman.

"Tell him to shut up," came shouts from the crowd.

Thumbs hooked under his suspenders, Gault continued to growl: "Hell of a dirty trick. Somethin' ought to be done. By God, a fella didn't always get a square deal. By God—" He remembered suddenly his speculation about what he'd do if the Company store refused him new credit; he suddenly felt that he would welcome such a situation.

The employment manager muttered to the doctor: "We'd better get out of here. Is he dead?"

The doctor replied: "Some chance."

"Take him to the hospital," Pendegrass ordered sharply.

The doctor didn't move. "He's a chance if we keep him here and the pulmotor going. He can't be taken out now."

The crowd went on bawling. The employment manager hesitated, his glance flickered up and his eyes scanned the angry faces of the workmen.

"I said: take him to the hospital!" he ordered, voice trembling.

The doctor looked up at Pendegrass: Pendegrass was boss. He stood up quickly and unscrewed the connection of the pulmotor from the light socket above the boardwalk. The employment manager knelt and helped him lift the body to the stretcher. Pendegrass' hands were white and quivering as he grasped the handles. The bluster had gone out of him; his clothes felt big on his body. He kept his eyes fixed on the pallid face of the body as the crowd let them through. He was afraid to look up. This was the sort of incident the Company didn't want to get noised about. He stumbled across the log-deck. Then angry crowd was behind him.

17.

The mill foreman began herding the men back to their places. The sorters and re-saw crews straggled up past the head-saws, but a few of the men still stood on the pond edge. Number One's sawyer said: "You sure told it to Pendegrass!"

Kurman smiled grimly.

"You sure told 'im," said someone else.

Gault growled: "He can't get away with that sort of stuff."

The crowd thinned more.

Kurtman said: "By God, it's time we was getting' together,—the workers."

Chins of the men jerked as if he had struck them. Shoulders grew rigid and eyes glanced furtively.

"We'll lose our jobs if we do," muttered Number One's sawyer, "look what happened to Blake in the factory."

But Number One's setter cut in: "Do it on the quiet!"

"We've a right!" said a dark-faced sorter, shaking a black fist.

"Well, it's time we did somethin', by God!"

"Ach, time!" growled Sheig.

"There'll be trouble over this. Take it from me, brother."

"If the kid don't pull through—"

Galt started up the slip, feeling a weakness in his bowels. This was not like old days—this unrest. He felt excited—afraid. Then he recalled the necessity of renewing credit at the Company store and again he grew strong just thinking how Becky and he would scrap if they refused him. He knew they wouldn't refuse, but his fear of trouble left him—he was ready for a fight.

On the pond edge, Kurtman and Sheig were almost alone and suddenly the feud over Marcella Condetta rose between them. Sheig's teeth bared a little, his eyes smoldering; then he turned and walked away. He would not give her up again—no punk like Kurtman would take her. He jabbed his pick at a huge log starting it toward the slip; then his eyes shifted to Kurtman going up the log deck. He thought to himself: ach, the stiffs shutt n't fight offer vimmens—they have too much else to fight.

Kurtman climbed to the carriage of Number Two. The power was on; the wheels rolled on the track; the frame trembled. He suddenly felt that the grudge between him and Sheig was stupid: he ought to do something about straightening it out. His glance met Gault's. He saw a think smile on the sawyer's face. Kurtman thought: guess Gus don't see himself so damned lucky. The saw whined into wood. He went on thinking: we'd better find out this noon what happened to that kid . . . The sound of the cutting rose to a frenzied shriek.

18.

Pendegrass took the steps of the administration building two at a time. "Nothing today—nothing today," he said to the waiting crowd. Inside he threw his hat on the desk and strode to his secretary's room. "Have you the card for the fellow, Lerner?" he asked.

"Was just going to send it to the accounting department," replied the girl, eyes startled. She picked up the yellow form card.

He took it. "I'll take care of this." His glance shifted as he added: "Forget about it."

At his desk again, he sat down heavily, laying the card before him. His fingers searched for a cigarette. He hoped suddenly that his wife would win the flower show—he felt that something must be done so that she would win. With a jerking movement, he loosened the trouser tightness over his bent knees.

The telephone rang and he took up the instrument.

"Hospital calling," a voice said. "Mr. Pendegrass?"

He grunted.

The voice continued. "Reporting on the man puilled out of the pond. He died five minutes ago."

Pendegrass coughed and put his cigarette in the agate tray.

"Do we dispose of the case as an employee of the Company?" asked the voice.

The employment manager spoke bluffly: "We've no record of the man here. Just a bum wandering around the pond and fell in. Turn the body over to the county."

The conversation ended. He tore the yellow card into bits, heaped them in the agate ash tray and applied a match. Again he took up his cigarette and smoked. Now if the millmen would just forget the incident—just forget it. In the old days, the men grumbled for a while over something like this; then thought no more of it. He recalled vividly the angry shouting of the crowd by the pond. Would they forget? A cloud of smoke volume from his lips as if knocked out of him by a blow on the chest. Would they forget?

The blast of the noon whistle shot upward into the fog-clear air, echoing over the roofs of the lumber mill village. The employment manager's fingers shook a little as he twisted the bright burning end of his cigarette among the black ashes in the green agate tray.

19.

The shriek of the head-saws died. Sheig had just brought another log to the deck; he hesitated before going back down the slip side. He seemed unaware that Kurtman had leaped to the deck near him. Neither spoke. They were again strange dogs sniffing.

The Kurtman said: "Still seeing Marcella?"

The pondman's fist squared out like a wooden block, gripping the shaft of his pick. "I marry wit her, maybe," he said.

They were confronting each other, standing firm.

Kurtman said: "She'll make you a good wife, Emil."

The tension was broken and they grinned at each other foolishly.

The setter from Number One came up. "Wonder how the kid's making out," he said.

Two sawyers had joined them.

"I'll find out this noon," said Kurtman. "Then we'll decide what to do."

"Ja. Goot," said Sheig.

Gault and the others nodded.

A sorter came past briskly, mumbling: "Foreman's coming."

The men strode down from the log-deck, silent now, faces determined. They did not look back at the motionless saws or the great logs on the carriages. They walked passed the spot on the pond where Lerner had been fished out not looking, not noticing, but from the set of their shoulders and the swing of their strides could be read no forgetting.[12]

11

PAUL'S WAR

The Depression naturalists saw life as one vast Chicago slaughterhouse, a guerilla war, a perpetual bombing raid.

  Alfred Kazin, On Native Grounds, *395*

One problem with a lot of rich people is that they are winners. They win and win, and winning becomes an obsession. It gets so important that the game itself doesn't matter. If they think they might not win, they rig the game.

  Richard Hugo, Death and the Good Life, *183–184*

Military Status

Jerre Mangione, a frequent visitor to Paul and Ruth's place at Cold Spring on Hudson, recalled that Corey studied developments in Europe and "was seldom at a loss to analyze the political and military strategy of any nation involved in the European war" and was certain that his knowledge would be of use to the intelligence service.[1]

Because he had been in the Reserve Officers Training Corps at the University of Iowa, Corey was commissioned a second lieutenant in the Army Reserve when he graduated. In 1937, with the rise of fascism across Europe, Corey inquired about his status. Army officials informed him they had discharged him on April 11, 1927, because they couldn't locate him and told him he could take the matter up with the local office, the Second Corps at Governors Island, New York.[2] He explained that he'd had no fixed residence after his graduation and expressed his desire to reenter the Army Reserve.[3]

He then began corresponding with the nascent local home guard, the Putnam County Defense Council, about working with them. The vice-chairman

DOI: 10.5876/9781646422081.c011

of the group wrote to Corey that he had suggested that Corey be in charge of implementing the program and assured Corey he would have a "good quota" of recruits in time for the next meeting Wednesday evening.[4] Corey was issued an identification card that described the unit:

> An Organization of CIVILIAN VOLUNTEERS formed to make available groups of men of the communities with some military training and the ability to act in unison under the direction of their own appointed leaders and with the approval of the constituted authorities, to provide prompt and effective means for meeting emergencies which may confront our communities in this time of National Preparedness for Defense.[5]

The day after the bombing of Pearl Harbor, the communist-affiliated League of American Writers issued an appeal for the United States to declare war on not only Japan but also Germany and urged writers to volunteer at the nearest Civilian Defense Volunteer Office. Corey received the call. With typical communist organizing efficiency appended to the entreaty was a list of places where members could sign up. Then, on December 19, 1941, Corey wrote to the Putnam County Defense Council that "someone had started a whispering campaign that I was an enrolled member of the Communist Party and that successfully sabotaged our efforts to build combat units for the civilian defense."[6] He categorically denied being a communist.

The vice-chairman of the group responded toward the end of 1941:

> I have your letter . . . and frankly it does not surprise me a bit. Although the reference to a whispering campaign is hardly accurate, I think I might as well tell you the facts and request that you regard this letter as confidential.
>
> The day after our original talk last summer, I was told by an influential and intelligent man in Cold Spring that I had made a great mistake, that you were an out and out Communist and were joining the Organization to act as a "cell."
>
> . . . . we tried hard to recruit a company for you but were met over and over again with the remark: "Why if anything happened, that is the first —— that we would shoot," and as you know, the thing fell through, in spite of our earnest efforts.
>
> Since the opening of hostilities I must have had 20 calls from various people in relation to you. . . .

As to rumors and all that sort of thing, you know that Cold Spring is a hot-bed and that as a rule they have little or no foundation and fall of their own weight. If there is anything that I can do, at any time, to assist you, please let me know and I shall be very glad to act.[7]

Corey responded:

> The "influential and intelligent man in Cold Spring" who started all this, if he is also a good American, should be willing to come before you, Commander Campbell and Executive Officer Eaton, and state his accusation and proofs in my presence, and give me opportunity of refutation. I [am] willing to take the judgment of you three gentlemen.[8]

At the same time, Corey began writing to every agency he could think of to ask for employment for the duration of the war. He started with the USO and proceeded to the National Defense Book Campaign and the Office of Facts and Figures. This became the Research and Analysis Branch of the Office of Strategic Services, which later became the CIA. The poet Archibald MacLeish of Yale and Harvard was director of the OFF.[9] Much later MacLeish would be instrumental in getting the last collection of Ruth's poetry published.[10]

Early the next year Corey was sent and returned an application form for the War Department, only to hear that no one was able to place him.[11] On April 29, 1942, he wrote in desperation to fellow Iowan Henry A. Wallace, Roosevelt's vice president. Wallace answered that the source of the difficulty was that there were already too many people in Washington and suggested Corey could be of greater service by continuing as a writer.[12] After being rejected by the Department of Agriculture, he wrote to Senator James M. Mead of New York and enumerated his applications, including trying to work as a carpenter at the Steward Airport, only to find that there was a surfeit of carpenters.[13] On March 28, 1943, he appealed to the president of the Farmers Union.[14] He was rejected by Columbia Broadcasting System.[15] The closest he came to getting a job was an invitation for an interview with a New York advertising firm.[16] In spite of these efforts, Paul remained, with one short exception, on the homestead for the duration.

In a similar situation was the better-known author John Steinbeck who wrote in 1943 that "everything in the government is so screwed up and complicated I am going to try private industry. It may not work but neither does the army nor the gov't. I ran up against nothing but jealousies, ambitions and red tape in Washington," that if he were in the army he would prefer to be a private as the rest was like the fraternity system at Stanford.[17] An East Coast club definitely ran the government, and Yale dominated the intelligence services.[18]

Home Guard

In December of 1941 Corey shifted his exertions from Putnam County to a wider arena. The New York left-wing daily *PM* wrote to him:

> Coordination of our civilian defense effort is admittedly one of the most important jobs in the United States today. It is our feeling on PM that the Office of Civilian Defense together with the Army, Navy and Air Corps are coordinating defense plans as efficiently and rapidly as possible.[19]

Corey then wrote to conservative magazine publisher Henry Luce to suggest that while it was reasonable to prepare for aerial bombing, the real danger was that a small number of planes carrying troops could land them, surround New York City, and cause immense damage and confusion before they were apprehended. He mentions that he had "hammered at Civilian Defense on this subject since August with no results."[20]

An editor at *Time* responded quickly,[21] but apparently nothing came of that. Toward the end of May 1942 a friend of Corey's, an editor at the competing magazine *Look* commissioned an article, based on Corey's synopsis, portraying an imagined airborne invasion of the United States by Hitler's forces.[22] Corey and others make frequent reference to a fifth column. The reference is from a fascist general as his four columns closed in on Madrid for the final battle of the Spanish Civil War, that his fifth column of supporters within the besieged city would undermine the defense and deliver the victory to him. The fear that fascists in the United States would organize and aid a German invasion was widespread and, at least to some extent, warranted.

With the payment of $250 (about $4,085 in 2019), the assignment would be lucrative.[23] The next month Corey wrote to Senator Mead to express his support for legislation to establish a Home Defense Force and outlined his efforts to arouse interest in such a measure:

In August, 1941, I attempted to organize Combat Units in the Civilian Defense at Cold Spring, but there was no authority for such a body and the local fifth and sixth columnists broke up the effort.

After December 6—

On Dec. 9 I wrote to PM describing the dangers. . . .

On Dec. 22 I wrote to TIME. . . .

On January 14, 1942 I wrote to the Office of Facts and Figures [Archibald MacLeish] stating again the danger of an attack by air. . . .

On February 22, 1942 I sent a brief on the danger of an attack by enemy airborne troops to the Office of Civilian Defense with arguments for a Home Defense Force and my offer to help activate such a force. (Carbon copy of brief attached.)

In March I approached an Editor of Look Magazine on this subject.

In early April LOOK Magazine asked me for an outline. . . .

On May 25, 1942 LOOK Magazine commissioned me to do this article. . . .

As you will readily see from the above I have done considerable work toward the breaking of ground for the activation of a Home Defense Force. I have received no official recognition or credit for this work. . . .

On January 16, 1942 I made application for re-instatement of my O.R.C. [Officer Reserve Corps] commission and passed my physical examination but have been unable to find out what has happened to my application since that time.

I am thirty-eight years old, am physically fit. . . . Yet I am not permitted to use my energy and ability to help my country at war. There seems to me to be something seriously wrong in such a situation.[24]

He wrote to his literary agent outlining his efforts and the *Look* commission and suggested, in addition, a small book and a movie.[25] At the same time he wrote to *Look* requesting a formal statement that he would retain all rights to the article aside from first serial publication and almost immediately received confirmation.[26] An editor wrote to verify the details of a meeting of Corey with the *Look* art director so the artist could see the locales covered in the Corey piece to inspire the illustrations that would accompany the text.[27]

The heavily illustrated 1942 article in *Look* opened with a single plane loaded with sixty-one ruthless, bloodthirsty Germans landing in a field five miles south of Fishkill, New York. First they cut the power lines beside the road. A milk tank-truck jackknifes across the road when they kill its driver. As the cars pile up behind the crashed truck, the Nazis drag the people from them and commandeer their cars to drive to key points where the commandos disrupt traffic and blow up bridges. New York City is cut off from its food supply. Darkness approaches. Police stop the cars, but the raiders take refuge in the abundant hilly wilderness. From this cover they continue to blast railroads, cut power and telephone lines, terrorize people, and ambush repair crews. "And the dread truth faces us: Virtually every industrial center or city in the United States is vulnerable to such an air-borne attack."[28]

Charles Sumner Bird and the Reactionary Crowd

Charles Sumner Bird could have stepped out of any of the chapters of sociologist C. Wright Mills's *Power Elite*: Celebrities, Very Rich, Chief Executives, Corporate Rich. Bird was the scion of an old American family, papermakers by trade, who early mechanized their production processes and expanded to all things having to do with paper. He was not from Paul Corey's usual circle of more or less bohemian writers, editors, critics, and neighbors. But he had seen the article Corey wrote in *Look* magazine on the possibilities of the invasion of the United States. On September 23, 1942, Bird wrote to Corey:

> What I want to have written is a story of what would happen if New England were invaded. This is to be used to show to important people in the Administration and our General Staff, how wide open we are. . . .
>
> There is a great opportunity in this story of showing what an essential force the State Guards are and how, if we are to be properly organized for defense, there must be appointed one man who has the responsibility for defense and in each area, a Commander under him who will work out plans for defense to coordinate all branches of the regular service and in the State Guards to cooperate as they are the only ones that can be everywhere at all times; it is the only force that can report enemy landing and delay action until regular forces can come. . . .
>
> Your "LOOK" story was splendid. The story that I want to get would bring it down specifically to an invasion of New England. Could you undertake it and if so, would you come on [*sic*] and go over the matter with me? I should

be very glad to pay expenses and I think in a short time we could decide on something that would be of great value. I know the situation in the War Department today and I want a story that will be a shock to them. . . . I am sure this is a matter of considerable importance.[29]

Corey answered:

I think you've given me a pretty good picture of what you want and I am heartily in agreement with you. My feeling is that we are sending our forces out to fight and have no adequate coordinated organization to defend our vital camp.

You have I feel, a greater fund of information on present defense organizations than I have, and we would need to talk the piece over thoroughly for me to get the exact slant you wish it to have. I've been hammering at our home defense weaknesses for over a year, but I've been sorely handicapped in my effort and therefore welcome the opportunity to help anyone who can drive home the issue to the right people. . . .

Unfortunately I will have to accept your offer to pay my expenses. I am not a very successful writer and we just manage to keep afloat. I would much prefer to contribute my time and travel to the war effort in this case until I can contribute greater service, but we haven't the wherewithal to do it.

In the circumstance I feel that you have the right to some assurance of my reliability. However, I know of no impartial source for you to contact. The editors of LOOK know me; there is my agent, McIntosh & Otis, Inc. 18 East 41st street, NYC, and my former publisher, the Bobbs-Merrill Co. 408 Fourth Ave., NYC, Archibald Ogden Editor. There is, of course, WHO'S WHO, but that can be of limited assistance.

In spite of the fact that we seem to be dreadfully alone in what we are doing, I feel that we are doing a vital national service and that something will come of it before it is too late.[30]

The Horror!

Bird paid Corey $250 to write the story, which he referred to as "Mr. Bundy's Horror Story."[31] It begins at midnight when rubber boats hiss onto the shingle of the Massachusetts shore and fifty grim Nazi cutthroats jump out to meet their Fifth Column guides on Horse Neck Beach. The silence of the shore and the sea absorb the sounds of the shots that kill any who challenge them.

The commandos cut the power and telephone lines on the New Bedford road. Heinkle 177 bombers from occupied France bomb Boston, Somerville, and Cambridge. Another dive-bombs Manhattan, and a third heads toward Pennsylvania Avenue in Washington, DC. An American fighter plane manages to shoot down the bomber, but it crashes into the Capitol after the crew bails out.

Meanwhile the marauders hijack cars, shoot air raid wardens, blow up bridges into New York, crash locomotives to disable rail junctions, and bomb docks. At 3:30 a.m. sharp the attackers return to their cars, and by 4:00 their rubber boats take them to the waiting submarine as it disappears below the surface of the sea. Dawn is near. Through the morning fog two of the new Dornier 217 E long-range bombers with the markings of Eastern Airlines land on an airfield in Connecticut. Raiders leap out to kill all with their submachine guns. Armed Fifth-Columnists help the Germans destroy manufacturing facilities. By 5:40 in the morning a heavy cloud of smoke billows over Hartford.

An intrepid fourteen-year-old grabs his bicycle and tries to warn various authorities. Each sends him to someone else. By the time members of the state guard are awake, dressed and assembled at their armory, the brigands are waiting and massacre them one by one.[32]

H. Wendell Endicott

The story in *Look* had caught the attention of another, perhaps even more potent member of the power elite, H. Wendell Endicott, who sent reprints of the story to a number of people. Endicott had written a handbook for the State Guard training school in Sturbridge, Massachusetts, that followed the first such school at Cambridge.

Late in 1942, at Bird's invitation, Corey took the train to Boston to meet with these two, and Endicott suggested a story that would include invasion by submarine in addition to airborne troops.[33] Corey went to Sturbridge to attend six days of classes at the First Service Command Tactical School, where he again met with Bird and Endicott.[34]

Corey arrived at the school in mid-October, when Endicott read the "horror story," and the three of them worked over the manuscript Corey had brought. Corey wrote to Bird:

Along with the script I am sending Mr. Endicott's notes. I quite agree that this must not be overdone, and I realize that in places I have laid [it] on too thick. However it is difficult for me to tell in all cases when that is true. It is also difficult for me to tell when my plot workings are correct because I know none of the three main settings of the conflict except in details acquired from books.

I would appreciate it to have the weak spots pointed [out] and suggestions as to the proper strengthening details. I'm sure you will understand that the enclosed script is merely a working draft, and that I consider it neither polished or finished.[35]

A Young Hero

The movie representative of Corey's literary agency began shopping Corey's movie scenario around.[36] He sent the scenario with the title of "American Guerrilla" to his agents in 1942.[37] Mavis McIntosh passed it on to Annie Laurie Williams, who began shopping it to movie makers. The New York office of Paramount Pictures saw the scenario, but there is no evidence that it was ever successful.[38]

The story was set close to a small suburban town near a port. The daughter of the owner of a war plant has a suitor who works in a garage and fears an airborne invasion. The plant owner disregards the young man, but the boyfriend organizes the members of a gun club into a militia. When a planeload of Germans crash-lands near the village, he is able to sound the alarm from a siren atop his garage, and the militia musters to block the roads. The raiders try to break through the roadblocks, and the girl gets caught between two barricades, but the resolute garage man rescues her, and his militia keeps open the lines of communication so that the raiders cannot surround the port.

The American People Can Fight

The solution for these multiple and diverse invasion threats? Clearly, a vigilante organization such as the one Otto Mantz got involved with in the projected fifth book of the Iowa work. The same kind of organization could slow down, if not halt, invading Nazis. Corey drafted an outline for a book,

"The American People Can Fight (Manual for Home Guard)," for Endicott and Bird.

It would open with motivation to join a guerrilla band, a reprint of Vice President Henry A. Wallace's stirring May 8, 1942, address to the Free World Association, "The Century of the Common Man," which would have been widely familiar at that time and appealed to fellow Iowan Paul Corey:

The search of the freedom—The march of freedom of the past 150 years has been a long-drawn-out people's revolution. In this Great Revolution of the people, there were the American Revolution of 1775, the French Revolution of 1792, the Latin-American revolutions of the Bolivarian era, the German Revolution of 1848, and the Russian Revolution of 1917. Each spoke for the common man in terms of blood on the battlefield. Some went to excess. But the significant thing is that the people groped their way to the light. More of them learned to think and work together.

The people's revolution aims at peace and not at violence, but if the rights of the common man are attacked, it unleashes the ferocity of the she-bear who has lost a cub. When the Nazi psychologists tell their master Hitler that we in the United States may be able to produce hundreds of thousands of planes, but that we have no will to fight, they are only fooling themselves and him. The truth is that when the rights of the American people are transgressed, as these rights have been transgressed, the American people will fight with a relentless fury which will drive the ancient Teutonic gods back cowering into their caves. The Götterdämmerung has come for Odin and his crew.

The people, in their millennial and revolutionary march toward manifesting here on earth the dignity that is in every human soul, hold as their credo the Four Freedoms enunciated by President Roosevelt in his message to Congress on January 6th, 1941. These four freedoms are the very core of the revolution for which the United Nations have taken their stand. We who live in the United States may think there is nothing very revolutionary about freedom of religion, freedom of expression, and freedom from fear—freedom from the secret police. But when we begin to think about the significance of freedom from want for the average man, then we know that the revolution of the past 150 years has not been completed, either here in the United States or any place else in the world. We know that this revolution can not stop until freedom from want has actually been attained.[39]

Speeches like this, with its praise of communist revolutions and labor unions, were likely the reason Wallace was removed from Roosevelt's ticket in favor of Harry Truman. The Dies committee in Congress was already actively hunting communists. Not everyone, as Corey was to learn, shared these Iowans' enthusiasm for the fourth freedom—the freedom from want—or their grassroots egalitarian radicalism.

The second section of Corey's proposed book would outline the need for a home guard in light of enemy objectives and our vulnerabilities. Third would be a section on details of organization, then weapons, and finally tactics.[40] Corey was not as alone as he thought.

Industrial Warfare

If any of this sounds farfetched, recall what the rest of the world looked like in the mid-1930s up to 1940. In 1935 the Italians conquered Ethiopia with bombs, poison gas, aerial strafing, and armored columns. Then in 1936 came Spain, testing ground for the German and Italian war machines. This was, significantly, not simply a war of Loyalists against Fascists. It was a war "of peasants against landowners, workers against bosses, anti-clericalists against Catholics, regionalists against centralists, and so on."[41] In 1939 the Spanish republic fell to the Fascist invaders.

The American, English, and French governments remained neutral. The Soviets supplied the Loyalists with weapons and military advisors. For the Fascists, the Germans and Italians provided air support, targeted bombings, artillery, tanks, and the personnel to operate and maintain them in this second test of mechanized warfare.

With the techniques of mechanized war worked out in Ethiopia and Spain, Germany proceeded to invade Poland in 1939; in 1940, Denmark, Norway, Belgium, the Netherlands, Luxemburg, and France; in 1941, Yugoslavia, Greece, and the Soviet Union. Paul Corey and Ruth Lechlitner watched reports of these events in the mass media from their place at Cold Spring on Hudson.

How could anyone respond to such mechanized warfare except by more mechanized warfare? One answer was provided by Tom Wintringham, commander of the British volunteer brigade in the Spanish Civil War in his introduction to the book on guerrilla warfare that "Yank" Levy wrote:

Those who believe that the tank is the dominant arm in modern warfare are correct. Those who see in the plane, and particularly in the dive-bomber, the decisive supporting arm that must be linked closely with the tank—are also correct. As a matter of fact, and not of theory, these weapons used in masses have secured decision in enough campaigns to establish the Nazi power from Madrid to near Moscow. . . .

. . . . Red armies of the Soviet Union have for months been resisting, without disaster, the forces of almost all Europe, organized with German thoroughness and flung into battle. . . . Their way of resistance has clearly proved effective and dangerous to Hitler's armies. And part of the method of the Red armies of the Soviet Union has been the use of the tactics of guerrilla warfare, not only by peasant snipers and bands of civilians turned guerrillas, but also by units, large or small, of Russian infantry surrounded by or passed by the swift . . . German armoured and unarmoured vehicles.[42]

He went on to argue that the weakness of mechanized warfare was in its long lines of supply for fuel, food, parts, ammunition, bombs, and the necessity to establish storage places safely behind the lines of combat. An armed and trained citizenry could wreak havoc on the logistics of such an army as the Soviets were doing. From the vantage point of 1940 Britain, it looked as though they would be next in line for Nazi conquest, and the British began to train their home guard.

Early in 1940 Levy had taught the first class in the United States on how to organize and train a home guard, as he had been doing in Britain with Wintringham. After the school was established, Levy returned to Britain. Levy's *Guerrilla Warfare* had been published through a joint partnership of Penguin publishers and *The Infantry Journal*, which marketed it in the United States as a cheap pamphlet for a mass audience. In the mass media people read stories of guerrilla warfare against the fascists in Yugoslavia, the Soviet Union, France, China, and the Philippines, so the pamphlet sold well, and local groups began to form across the United States.[43]

Guerrilla Response

From the school in Sturbridge, Corey wrote to Ruth back in Cold Spring that the food was good but that it was difficult to sleep because no one had turned down the thermostat. He wasn't official, he wrote, but could observe where he wanted and do what he liked. Endicott was hazy about "the whole thing," and Bird had asked to read his horror story. "I don't know what the report will be," he wrote to Ruth, "I suspect that it's one of those things which depends upon what worm the Bird has eaten before he reads it."[44]

Corey made a number of suggestions that the colonel in charge of the school liked.[45] Three days later, Paul wrote to Ruth and said he would be on maneuvers in Rhode Island and Connecticut on Sunday and Monday would go with the colonel to Boston to meet Bird. "I won't know the status of the horror story until then apparently."

The school had finished, and Corey found it more advanced than anything else for understanding offense and defense. "I've been living in a great deal of uncertainty about a great many things up here, which is apparently the Army way, but I hope I can clear a few of them up before I come back."[46]

Bird wrote to Corey at the school to meet him at the Somerset Club Monday evening and that he would reserve a room for him there so they could talk.[47]

Boston's Somerset Club was a drastic change of scene from the farm at Cold Spring. This club was and remains a center for New England's traditional elite. During the Civil War, when members were not unanimous in their support of the United States, a separate Union club split from it, leaving the rest who supported the Confederacy. In 1945, three years after Corey was there, when the club was on fire, the responding fire fighters were invited to use the service entrance.

Five days later, Corey sent Bird a table of contents for the book *Our Citizen Army*, with the promise of getting a draft done within a month or six weeks, though he said he couldn't assess the chances of its being published. He explained that he would want an advance from a publisher of $500 to write such a book but that he would do it for Bird for $250 with the promise of being reimbursed if the book were published; he promised to send a final draft of the "horror story" by November 3 incorporating changes suggested by Bird and Endicott.[48]

Bird responded that the outline was "pretty thorough" and that he was looking forward to the "Bundy," story, the "horror story."[49] The next day he wrote to urge Corey to go ahead with the book, in somewhat simplified form with some other specific comments.[50] Corey responded that he was under the impression that the book should be patterned after the British prototype,

Home Guard for Victory by Hugh Slater, and argued for the necessity to focus on a narrow range of topics. He added, "I have disciplined myself to selecting one purpose at a time and discarding everything that doesn't bear directly upon that purpose." He enclosed an outline.[51] The next day he elaborated, "I tried to build a structurally well-proportioned book with the unified purpose of showing the State Guard and their job."[52]

Bird sent Corey the manual that Wendell Endicott had recently published, including a number of the lectures from the school Corey had attended. He continued that a short book would be most effective and that he and Colonel Green (of *The Infantry Journal*) thought that a book about the size of Levy's *Guerrilla Warfare* would be useful.[53]

The Gospel of Wealth

This correspondence begins to show the fundamental difference in outlook between Bird and Corey. Bird was born to incalculable wealth, and Corey's inheritance was work inconceivable to Bird. Corey's new outline starts with chapter 1: "What We Are Fighting For." The first topic in the outline is "The Enemy We Fight," the second is "The United Nations Charter," and the third is "The Century of the Common Man." The second section on civilian armies lists England, Russia, and China as examples.[54] "The United Nations Charter" would include the four rights Roosevelt articulated in his 1941 State of the Union address and that Vice President Wallace had mentioned in his speech. These were freedom of speech and worship, already guaranteed in the Bill of Rights, but to them he added freedom from want and freedom from fear.

Bird wrote to inquire again how much Corey would charge for the work of writing the book.[55] Meanwhile, Corey wrote that he'd read Endicott's manuscript of a guide book for home guards and found that the project he and Bird were discussing would be a waste of time, as it would be redundant and reach only a small audience. Rather, he suggested a "vicarious Dunkirk":

> If we had five million State Guardsmen in the eastern states, those two books
> [Endicott's and Bird's proposed book] would prepare them to meet and
> defeat any raid the Nazi could send against us, but with the few thousand of
> State Guardsmen we now have such books can do little good.

Why don't we turn all our efforts to breaking the War Department's reticence on arousing the public to the danger which faces it? Then we'll arouse the public and when the public is aroused, we'll give it Mr. Endicott's book and yours to show the public what to do in the situation.

That seems to me to be the logical way to get this job done efficiently and quickly.[56]

The next day he wrote that he could do the book in four to six weeks because much of it was already done, that it would be of little value to him as an author but that he could do it "on a straight fifty dollar a week basis." But, he concluded, with the current state of public apathy he doubted they could even find a publisher for the book.[57] In spite of Bird's apprehensions about civic indifference, the enthusiasm for such a home guard in the United States was uneven; there were some enthusiasts, some opponents, and most didn't care one way or another as long as it didn't bother them. Corey summarized that he could do a pamphlet incorporating the preexisting materials Bird wanted:

> I'm willing to take the material which you gave me and smooth and polish
> it up for you at fifty dollars a week. I can probably do it in four weeks or less,
> depending upon how easy the re-writing goes. Then of course you can do
> what ever [sic] you wish with it.[58]

Bird responded that he wanted the book to have some popular demand rather than be limited to training. He continued, "Chapter 1—'The United Nations Charter'—I should leave this out and replace with the Bill-of-Rights." He continued with suggestions for Corey's outline and asked again about the price and time.[59] Corey answered that the more compete book would take longer than the pamphlet they had been discussing but that he would do it for $250; he returned the outline for any further comments but asked, "If you want the Bill of Rights substituted for the United Nations Charter, do you still want 'The Century of the Common Man' included?"[60]

Bird replied that the editor of *The Infantry Journal* was interested in bringing out their book and continued:

> I should think that the Constitution and the ten amendments known as the
> Bill-of-Rights, were the most important document that the common man has
> ever established. Without it there would be no opportunity whatsoever for

the common man. It is the protection of the common man and the power of government which always leads to some form of tyranny.

I see no objection to leaving "The Century of the Common Man" in the first chapter but I am very strongly of the opinion that in the United Nations Charter there also should be added a fifth freedom—that of the freedom of the individual and free enterprise, otherwise, we shall be turning backward rather than forward. Throughout the ages the progress of mankind has been measured by the development of men with character and quality who appreciate the fact that man, the individual, is the measure of all things.[61]

Here Bird quite naturally echoes Andrew Carnegie's gospel of wealth and the rhetoric that the du Pont–backed conservative anti-New Deal American Liberty League was providing.[62]

Corey responded:

Personally, if I were assured of the four freedoms, I'd need no more guarantees for me to realize my individual expression as a good citizen. So what is the necessity of a "fifth Freedom"?

Frankly I don't understand the meaning you have for "Free Enterprise." If we assume absolute and universal freedom of enterprise, we will have, in a very short time, the stronger curtailing the freedom of the weaker and there would be no longer any freedom of enterprise. A man with a thousand dollars has more freedom of enterprise than a man with ten. If on the other hand we set up rules to keep the stronger from curtailing the freedom of the weaker, then we don't start with freedom of enterprise.

It seems to me that up until about a year ago Hitler had the greatest freedom of enterprise of anyone in the world and we know what he did with it. If everyone could be depended upon to be a good citizen, then freedom of enterprise would be safe; but if everyone could be depended upon to be a good citizen we wouldn't even need any laws. So what meaning do you wish the phrase "free enterprise" to have?[63]

Corey didn't take naturally to the gospel of wealth, but Bird was quite willing to explain:

"Free enterprise" means the opportunity for man to use the best that is in him without restrictions imposed by government when the actions are in no pay [*sic*] injurious to the commonweal.

Communism, Socialism, Fascism, Nazism are all the same sort of philosophy which means dictatorship. The four freedoms in the Atlantic Charter are great ideals but it is totally impossible for a weakling and evil man to be free from want. If you get into a discussion of this situation you get involved in a discussion of a social character which must be avoided.

If you do not understand the meaning of "free enterprise" I do not care to go into it and discuss it and if you deal with it too much in this book, the book would serve no purpose whatever.

Whatever you may think of the United States, our society has been developed under free institutions. The State, by organization and machinery, when powerful, destroys the individual and all sense of brotherhood.[64]

This was an echo of the American Liberty League's rhetoric. Corey responded that he agreed that there shouldn't be anything controversial in the book and that, with all his questions answered, he was ready to proceed whenever Bird said to.[65] Bird asked for a final table of contents and estimate of costs supposing it would be published by *The Infantry Journal*.[66]

Corey sent a revised table of contents with the title of "Our Civilian Army." Chapter 1 was devoted to the Germans, Japanese, and Italians and chapter 2 to "What We Fight For," which included the Bill of Rights, the Four Freedoms and a fifth, and the Century of the Common Man.[67] Bird accepted the revised table of contents and suggested Corey proceed, though he stated he didn't know whether it would be sufficiently interesting for *The Infantry Journal*.[68] By early March 1943, Corey could report that he was nearly done, and on March 9 he sent Bird the manuscript.[69]

The Gospel of Work

These experiences with the upper crust must have informed Corey's writing of his 1946 children's book, *The Little Jeep*, the story of Joey, a little jeep that was tired of combat and wanted to return home. Because the story illustrates the gospel of work, I will summarize it here.

At the beginning of the war, Joey had wanted to be a tank, but the tanks laughed at him and went clanking about their business of smashing people and things. But now that he was headed home, he wanted to do something useful. He learned he wasn't cut out to be a ship like the one that volunteered to take him home. At the dock after the voyage, the ship advised Joey to find a useful job as soon as possible. When he approached them, the taxis shouted at him angrily and laughed at him. He raced away to streets lined with fine houses with haughty, shiny black limousines in front. One of the limousines questions Joey:

"What are you doing on my street?" he asked gruffly.

"Is this *your* street?" replied Joey. "I thought streets belonged to everyone."

"I carry a rich man," said the big limousine. "He likes the street to look nice. He doesn't want any dusty, scratched jeep in front of his house. So get out of this street right away!"[70]

When he came to a railroad yard and saw the engines, he thought, "There's a real job." The engines conferred and told Joey to go to the big streamliner and offer to pull his train. *"You!"* snorted the big streamliner. "You insulting little piece of junk! Those mean, jealous freight engines put you up to this!"[71] The freight engines laughed as the little jeep hightailed it out of the yard. Determined to find useful work, he continued to the airport and noticed how beautiful airplanes were when they weren't dropping bombs or strafing.

But he could not fly because he had no wings, and a transport plane said, "Oh, don't bother me with such silly chatter, I've got to take off and fly three thousand miles to New York." The jeep left the city and ran across the country beyond. He came to a junkyard where he saw tanks and cannons and the machines he'd known in the war. He recalled how strong and proud they had been, how much they had destroyed, and now they weren't good for anything because they could only destroy things.

He thought perhaps he should join the others in the junkyard, but at the top of a hill he saw fields waiting to be plowed and planted and a worn-out old tractor coughing and grumbling and sobbing. "The land must be plowed and planted so that the world can have food, and I'm too tired and old to do the job."[72] The farmer saw the expired tractor and agreed to give the jeep a try. "And as he worked, his engine purred, his hood shone, and he seemed to smile all over."[73] The farmer stopped the jeep and called to his wife: "Come and look at how much work a little jeep has done. Now we'll get our fields plowed to raise food for the world."[74] He patted the little jeep's hood once more and stroked his fenders. Joey started pulling the plow again, happy to have found useful work.

Militias

But the war was not over yet and the future was in the balance. Corey's various scenarios for the invasion of the United States and several of his correspondents mention, as though quite natural, a fifth column. The term was widely used in the press in the United States with respect to the widespread fascist movement in the United States.

Corey was clearly frustrated by the ambivalent reception of his views on the necessity for an armed vigilante movement to meet any threat of outside aggression, but he seems to have been unaware of the role of such organizations in the rise of fascism in Germany and Italy. Official ambivalence toward the matter may have been conditioned by this wider view of the possible roles of such groups of armed citizens.

The ideas were not unique to Corey and Bird by any means. In response to his *Look* piece, the magazine received number of letters from readers addressed to Corey at the magazine's office in New York City.[75] They sent at least some of them on to Corey. One was a letter from Columbus, Ohio, that described a group of 200 active men with officers, uniformed in steel-gray outfits, eager to get going and maintaining regular drills twice a month.[76] The head of a township group from Riverhead, New York, mentioned that in New York such organizations had to be deputized as deputy sheriffs and that members of his organization were working people who provided their own cars, gas, and equipment. Armed with high-powered rifles and plenty of ammunition in addition to shotguns, they stood prepared to repel any invasion or deal with any sabotage of a fifth column.[77] A September 4, 1942, newspaper article in the *Cleveland Press*, titled "Seeks Volunteers for Guerrilla Band," reported the formation of a guerrilla battalion in Cleveland with a hundred candidates from Republic Steel Company under the leadership of a member of the plant protection department.[78] A member of Company 99 of the Virginia Reserve Militia reported that the National Rifle Association had suggested a plan for a Home Guard and that most states had reserve militias organized and ready to go into action.[79] A member of the Second Missouri Infantry of the Missouri Home Guard wrote that they were the first such unit organized, the first to be in uniform, and the first to engage in firing maneuvers.[80] So, according to these letters, there were Home Guard units already organized in Columbus and Cleveland, Ohio, Riverhead New York, Cambridge Massachusetts,[81] Barren

County, Michigan,[82] Virginia and Missouri as well as the Pacific Coast Militia Rangers in Canada's British Columbia.

There's a letter from "Karl" at *Look* dated September 9, 1942, saying, "I am sending you this one because it's an unusually sincere and intelligent letter. I think the lad would get a lot of satisfaction of an answer directly from you. I don't intend to keep pestering you with these letters, of course. We're getting a nice response on the article. I can keep all the letters together, if you like, so that you can have a look at them when you get to town one of the days." Corey must have known Karl because Karl sends regards to Ruth in another note, but I have not identified him further.[83]

American Fascism

After a decade or more of foment and formation of radical groups on the left and the right, not everyone was as enthusiastic about armed militia units as Corey and Bird. There was the example of the use of such organized groups in the rise of the fascist regimes in Germany and Italy. But there was also a similar history in the United States, at least close enough to make people anxious at the idea. In 1939 there were 800 fascist groups in the United States.[84] Perhaps the best known is the plot exposed by Major General Smedley Butler of the Marines.

In 1933 after four years of worsening depression and Hoover banalities, the American people ousted the pontificating president in favor of Franklin D. Roosevelt and his program of action known as the New Deal to repair the economic system. "Repair" is a relative word. To some, the economic system was not broken or even malfunctioning—it was keeping them rich.[85] But to the half of working people who were out of work; to the farmers who were losing their farms to repossession; to the vast majority of Americans, the economic system was the cause of their misery. That it worked to benefit of the very few did nothing to ameliorate the misery of the many. So they voted repeatedly for Roosevelt until he died before the end of the Second World War in 1945.

Roosevelt's program may not have ended stock speculation, but it curtailed it to the consternation of the bankers and financiers. In addition, it stopped farm foreclosures. It regulated telephone, telegraph, and radio industries. It provided a legal framework for labor unions that prevented employers from hindering union organization and compelled management to accept collective bargaining.[86]

These moves spawned immediate and intense opposition by that small minority whose rapacity was being reigned in to benefit the great majority that voted. In her 1939 tour of the United States as the economy was beginning to respond to these programs, journalist Anna Louise Strong documented internal and external sabotage of the New Deal at every level from local to national.[87]

Much of the familiar political rhetoric of the first decades of the twenty-first century echoes the strident tone of the organized opposition to the New Deal that started upon its inception, and many of the propaganda techniques were developed at that time when the potential power of the mass media of newspapers, magazines, and radio was just becoming apparent. Then, as now:

> What America really required "was a drastic transfer of wealth from the
> Haves to the Have-nots." As yet no one knew whether the Haves would sur-
> render their power and possessions without a struggle, or whether Roosevelt
> was prepared to battle them if they refused.[88]

In those days, the economic elite could organize themselves more transparently than today. The Liberty League was perhaps the largest and most vocal such organization backed by the du Pont family and other industrial and banking giants.[89] A member of that minority with uncharacteristic frankness wrote:

> My kind, in my time, grew up to order, security and privilege of a degree
> which it is difficult for most Americans to gasp. Their world-view, ethics,
> and code of manners, all their mechanisms, were aimed at the perpetuation
> of that security. Now [1945] it is more than threatened, it is in the core of
> being destroyed. This is an old situation, recurring over and over again in the
> world's history, it's a situation which arouses unexpected viciousness in those
> whose way of life is threatened.[90]

It is certain that Roosevelt's agricultural program was intended to quell rural unrest and was successful at doing so.[91] The New Deal envisioned three coequal branches of the American economy: industry, agriculture, and labor. Industry was already organized; New Deal laws allowed labor to organize, and the farm policy was to shape agriculture into a single industry whose interests would be represented by the Farm Bureau.[92] Some suggest that his New Deal was meant to "choke off the kind of lower-class discontent that had produced the Hitlers and Mussolinis in Europe."[93] But it was the great industrialists who were at the center of European fascism. The lower-classes of Italy and Germany were socialists and communists. Fascists wiped out those sources of opposition early on.

The Dies committee took the first steps toward the Great Repression in the United States but didn't complete the job until the 1950s. Henry Luce, publisher of *Fortune* and *Life*, devoted the July 1934 issue of *Fortune* to the glorification of Italian fascism, praising its preservation of ancient virtues such as discipline, duty, glory, and sacrifice.[94] It was this same Henry Luce to whom Paul Corey wrote to pitch a story on the necessity to organize and arm a home guard to protect against Nazi invasion threats.[95] "In the summer of 1934 [in the United States] it was not difficult to detect the acrid smell of incipient fascism in the corporate air."[96]

The American Liberty League was openly affiliated with fascist organizations in the United States.[97] If the inevitability of a communist revolution was apparent to some, it was also taken seriously by the industrial elite. State militias similar to the one Corey was contemplating were trained to confront mass demonstrations. Scores of protesters died, and thousands were wounded at the hands of these militias across the country.[98] Anna Louise Strong provided firsthand testament to such events in Minneapolis.[99]

In 1934 Huey Long mobilized his Louisiana state militia to neutralize the powerful local political machine with which the mayor worked.[100] With its unlimited wealth, the American Liberty League not only orchestrated a massive propaganda program against the New Deal and Roosevelt; it supported any and all other opponents, including Huey Long.[101]

In Philadelphia, Arthur J. Smith founded the Khaki Shirts modeled on Hitler's Brown Shirts and Mussolini's Black Shirts.[102] They were recruited from former military men, many of whom had participated in the Bonus March that Douglas Mac Arthur had dispersed. Smith espoused fascism and claimed to have organized half a million men who were prepared to take control of Washington, remove Roosevelt, and establish a dictatorship as an escape from economic injustice and global depression. Roosevelt appointed J. Edgar Hoover to investigate the twin threats of communism and fascism.[103]

Against this backdrop of well-organized and well-financed fascists in the United States, when Smedley Butler, a highly decorated, experienced, and respected retired Marine general, announced that he had been approached by a coalition of businessmen with a plot to seize the government, he was taken seriously.[104]

The House Committee on Un-American Activities was organized to investigate such organizations and heard Butler's testimony. The press, led by the Hearst papers and Luce's magazines, including *Fortune*, at the insistent urging of the American Liberty League, derided his revelations of meetings with representatives of industrialists and bankers to recruit him to lead a coup d'etat. But those who discounted his disclosures were the same people who had approached him.[105]

Nazis and Communists

In an undated handwritten page in the University of Iowa Library's file containing the Bird-Corey correspondence is this note, presumably by Corey:

> We were slow to believe that the Nazi could be bestial, because the Nazi told us they were the bulwark against Communism and Bolshevism. And the Profascist Shouter told us that the Bolsheviks and Communists would take everything we had and therefore we believed easily that they were beasts and murderers. But the Nazi have robbed and raped and murdered and destroyed two-thirds of Europe, and the Russian Communist Regime has led 188 million Russians in a magnificent stand against the Nazi—a stand which has given us time to prepare to save ourselves. The Bolshevik beast and the Nazi Savior has turned into the Nazi Beast and the Bolshevik Saviors. And we should not forget that in time to come—when some are again out to save us from the Communists.

The Dies committee was the harbinger of those times to come. The House Un-American Activities Committee was no longer interested in fascists. And the FBI had never heard of the four rights that Roosevelt wanted to guarantee.

12

RUTH'S WAR

... if an artist wants to marry he should find a woman who doesn't mind
doing her own housework.

Oliver La Farge, Raw Material, *209*

Left Out

Built into our language from ancient words is the idea of the man protecting the woman, the husband and wife. But Paul and Ruth reversed that pattern. While Paul Corey turned his attention from Iowa farms to the lurking
German menace, Ruth continued to work and write. Her struggle was to
see her work beyond individual poems in *Poetry* and *New Masses*, little magazines and slicks, to make it available to a mass market in books and to build
a national or even wider audience as Robert Frost had done.[1] While it took
Corey a decade to find a publisher for his trilogy, editors were asking Ruth
for her poetry. Finding an outlet wasn't the problem. Finding a market was.

In 1940 Paul and Ruth sold the original farm building they had been living
in and increased the chicken flock so they could sell more eggs.[2] In September,
having finished building the new house with electricity, running water, a
kerosene stove, and a washing machine, Paul decided their circumstances
had improved sufficiently to support a child. According to their friend Jerre
Mangione, Paul was certain he would get at least a reserve commission and

DOI: 10.5876/9781646422081.c012

did not want to leave a lonely wife behind. Therefore, he waived his objection to having children so that—by his way of thinking—Ruth could occupy herself with the child while he was away.[3]

Not inherent in the communist program was women's freedom to "exercise their infinitely varied gifts in infinitely varied ways, instead of being destined by the accident of their sex to one field of activity—house work and child-raising."[4] While Ruth's domestic work may sometimes have been more akin to drudgery than most, it was not hers to bear alone, nor was she burdened with child-rearing until relatively late in life, and then surely at her own behest and with much support and assistance.

The writer Oliver La Farge was from an eastern family of such stature that they did not do their own housework. Ruth seems to have done not only her fair share of housework, cooking, cleaning, ironing and such but also to have provided the cash income for the household until near the end of her life. In the view of La Farge, writers' wives

> have to have enough of the artist in themselves, plus enough character, to be genuinely more interested in the production of good work by their husbands than in comfort or security. It takes more than a little not only to accept but to further the man's decision that he must abandon a profitable but exhausted line; some women can't take it, those who can are the kind who don't just take it, but see the reasons and are all for it. When a man is seized by as cockeyed an idea as writing this book, she does not stop to consider that it's not likely to make money or that the project involves telling things about himself which will cause some people to go "tchk, tchk!" She considers first and exclusively whether in her opinion it will produce good writing, and if she thinks it will, makes preparations to alter year before last's dress again and urges her husband on. In return for which she gets his excessive mood-swing, his irregular habits, chronic trouble over bills and lots of practice washing dishes. Women like that don't grow on every bush. I like artists, but their wives I find magnificent.[5]

Third Pregnancy and First Poetry Book

In any case, Ruth got pregnant immediately. With the threat of abortion removed she was full of energy and hope.[6] After a seven-and-a-half-hour trouble-free labor, their daughter was born on May 26, 1942, in a hospital.

Her doctor did not arrive in time because he was convinced that no woman of forty could have a short period of labor. Anne, the new daughter, took immediately to breast feeding, and Ruth nursed her for four months before returning to work and leaving the baby with Paul and a bottle.

Paul was writing a novel in the mornings and growing their food the rest of the day. Leaving the baby with him would be routine for him and allow Ruth to bring in some wages. After a day's work in the city, she would ride the train to Cold Spring Station where Paul and Anne would be waiting among the mothers and children awaiting their men. Ruth surmised that she must have seemed to Anne more like a father returning with the other men than a mother who stayed home with the children.[7]

So it is clear that Ruth was relieved of many of the duties of childcare and equally clear that the income from her New York job sustained the family. But she continued to write and publish poetry. At the invitation of Willard Maas,[8] her first book of poetry was *Tomorrow's Phoenix* (1937) from Alcestis Press, which by this time had published eight limited-edition volumes by Wallace Stevens, William Carlos Williams, Allen Tate, Willard Mass, and Robert Penn Warren. In 1938 the press folded and its director, the enigmatic Ronald Latimer, moved on to other things.[9]

Lechlitner's volume was devoted to revolutionary poetry. Harvard-educated poet Samuel French Morse praised it among the "recent outpouring of Left-Wing literature,"[10] for which standards of criticism had been relaxed because of the importance of the subject matter.[11] He applauded her lack of dogma and the integrity and power of her verses.

References to the League of Nations and the civil wars in Spain and China date Leichlitner's 1938 poem, "Ordeal by Tension." But surely the references to the "steel hawks against moonlight," threats, pacts, ambassadors and dictators, hungry unemployed young men with no alternative to military service, the flight of reason, compromise, memory, and the building tension whose only relief is war—these themes ring as true in the second decade of the twenty-first century as they did nearly a hundred years before.

Ordeal by Tension

> *The fanged love of nations,*
> *The fisted enterprise compelling*
> *Steel hawks against moonlight*

Over nocturnal Europe
Knowing these, our danger's
Not when the wings plunge, but now:
Now when the tongue's sentry
Sleeps, and the enemy in the heel
Itches for action,
Now when flesh suffers
The ague of suspense:
Ordeal by tension

The drums beat, the drums beat—
We hear them in stale rooms,
We lean from the windows
Listening. Only the empty street.
(No banners, no sharp glint
Of sun striking metal, no marching.
We dreamed, then. Mock, dream, the quick
Pulse leaping!)—Here's daylight,
Accustomed chair, breakfast, morning paper.
But again the drums beat
In the scare-bloated headlines:
Threats, pacts, empty treaties,
Ambassadors, dictators,
Blood-darkened bargainers
In the world's market-place.

This is our waking; and waiting
Helpless while drums beat, morning
By morning, puts day out of focus:
Walking the perilous
Tightrope of tension
Earth sways beneath us,
Lost balance betrays us.

These signs of our sick time, contagion
Of sanity broken,
We carry from house to house
(Death waits in those houses)

In subway and restaurant,
Shop, theater; in the doomed faces
Of strangers we find them,
In talk between friends:

"One move, and they're ready."
"The League's scrapped and mortgaged."
"In Spain now—and China . . ."
"Can't hold out much longer:
Christ, then, let's have it!"
"Build up our battleships:
. . . then we'll be in it!"

The drums beat, the drums beat
They call to the young men
(The young so long bedded with
Idleness, hunger)
What other life for them?
How shall we censure them?
We too are proud,
Keyed to adventure
To leap the first barrier:
How well they know us—the traders
In men and profit, the makers
Of skilled propaganda!
They harden and test us

Like fine-tempered metal:
Shivering by day and sweating by night
We flush at each rumor,
Rebel at retraction, Cheer men at the barricades,
Cry for defense.
(Gone memory, honor,
Compromise, pity,
And reason, that poor ghost of peace)
Steel whistle, blood music:
The drums beat, the drums beat
Till we shout for war—the only release![12]

The Second World War has receded into history and resides in the national memory alongside the Civil War and the First World War. Writing on the eve of the Second World War, seven years after the publication of Ruth's poem, La Farge conveys the sense of the developments he was seeing:

> This catastrophe came upon us so slowly, remorselessly, and visibly that to varying degrees its imminence distorted us all. Essentially, we lost our assurance of security in 1929. It was not so long after that that we knew what was coming next, although we used many devices to deny it to ourselves. Expectation grew tenser year after year; the tension affected all our thoughts. It pushed some people into downright hysteria and caused us all to give serious audience on occasion to lunatics. The method and immediate timing of the attack on Pearl Harbor were a surprise . . . but the fact of War at last was not. It was a great relief. The catastrophe had arrived after our long waiting, we could relieve our souls with action.[13]

Answers

The global tensions mount with the rise of European fascism, and then just half a year before the bombs drop on Pearl Harbor, Lechlitner appropriates from radio the popular form of the quiz program to query her readers. She speaks of lost jobs, enemies, allies and small nations, embargos, and the House Unamerican Activities Committee gnawing at democracy's bone. "Bewildered," she writes those long decades ago, "we turn our ears / To all the great networks of the screaming air" to answer why we are afraid and to ask what name time may have for these black years.[14] These years later we know that one—the Second World War and the Greatest Generation. But Americans are asking the same questions today as we leave the second decade of the twenty-first century. Perhaps they are endemic in the warfare state. Once again we find ourselves not knowing the answers, only the "bare shape of silence among us after we turn off the media shouting." The "fourth term" refers to Roosevelt's fourth term as president before there were term limits on that office.

> *Whatever happens, let us dedicate this hour of our lives*
> *To question and answer: tomorrow may not roll on little wheels*
> *Into our bedroom, the future may be now, the far-off here,*

We may neither guess the time that lies under our heels
During the last pretense of sleep, nor dream
A real dream in daylight: let us fear
Nothing for an hour at least—maybe we'll know and maybe
win the prize.

This one comes from a man in Lexington, Ky., asking who
Stole his job . . . but he doesn't say when, so
Let's take another. Speaking of enemies and allies, why
Are the sins of the small nations unforgivable? No, lady, no.
Because taxi-drivers are seldom oracular,
With grief like a bandage around our heads, we try
Not to think too much before breakfast. True or false? True.

Listen carefully: this is harder. What followed when
Three German aviators were dropped on an English shore
(Green waves took them flaming, death grinned from their eyes)
—We lifted the Embargo because there weren't four.
But speech is not always free, and if as you say
The dog gnawing Democracy's bone is Mister Dies
Whoever believes that dew falls can still believe in the rights
of man.
You must guess two out of three: "Leave now the lovely light"
Is from what poem; when marketing will you buy
(Wheat and corn steady, steel closing up fractions, aviation firm)
Red plum or white cauliflower or the shattered thigh
Of a woman bombed, or the torn hands of children . . .
If love overcomes all, will there be a fourth term?
Sorry: you lose. Be with us next week, and thank you: goodnight.

Desperate for the sign, the clue, or the word, our eyes comb
The headlines and bylines. Bewildered, we turn our ears
To all the great networks of the screaming air.
Why are we afraid? What name does Time have for the
black years
Marching before us? Tell us, Professor Quiz and Information
Please,

We don't know the answers. We know only the bare
Shape of silence among us, after the Voice in the room[15]

The poem ends without a period.

Columbia Broadcasting System's 1937 broadcast of Archibald MacLeish's verse radio play, "The Fall of the City," narrated by the young Orson Welles, was a broadcasting sensation. A distinguished roster of American poets—such as Norman Corwin, Edna St. Vincent Millay, Alfred Kreymborg, Steven Vincent Benet, and Carl Sandburg—followed. Among them was Ruth Lechlitner with "We Are the Rising Wing" in 1938 and "Tale of a World's End" in 1957.[16]

Disappearing Decker Press

In the United States the process of repression of radical poetry has been as successful as the disappearance of other radical writing. Ruth continued to write poetry nearly to the end of her life, even though 1939 had been announced as the death knell of radical poetry. Like other radical poets, her work appeared in "tiny press editions and alternative magazines" and was largely ignored except by those in her circle of radical writers and editors.[17]

It may have seemed easy for Ruth to get a collection published in book form. After all, Maas had invited her to submit her first volume to Alcestis Press. The fact that the press issued very limited editions of collector volumes at very high prices was not in itself a death knell. Others the press published became well known. But it was no entrée to the world of poetry publishing if for no other reason than the transience of the press and its owner.

Ruth also received an invitation to submit a collection to the Decker Press, founded by James A. Decker, but the press's acceptance of her work in 1942 only started a frustrating two-year correspondence. A single letter is suggestive of the rest:

> While I sympathize with your aims and ambitions for the Press, I am afraid I cannot send you $20 or more to help with this advertising [in the *Saturday Review of Literature*]. My book, Only the Years, was contracted for early in 1942; nothing was done toward its publication till late in 1944 when I sent the

Decker Press $160 to pay for the binding costs—or the volume would not
have been published at all. Then, in February 1945, the book appeared without a jacket (other Decker volumes I had reviewed were issued with jackets).
To shake further my confidence in the Press, review copies were not received
by periodicals that should have had them, including *Poetry*. Up to the middle
of December 1945 the latter still had received no review copy, as Leo Kennedy
ascertained when he called Miss Peters on the phone, and sent over to her his
own copy of the book that they might review it.

I realize that, during the War years, you worked under great difficulties;
at the same time certain authors were published by you without delay, and
in handsomely bound, jacketed editions. I can only assume they were able to
pay more than I was. . . .

No one wishes more than I to see a press like yours free, independent, and
self-sustaining. But I wish its continued existence could be insured by patrons
or sponsors—not by the poets who have manuscripts to publish. If the
Press—the publication of books of verse—is your only means of livelihood,
then frankly I think you would do better to become strictly a "vanity" house.
You cannot be half one thing, half another, without hurting your Press and
the better poets you publish.[18]

Three months later Alfred Kreymborg wrote to Ruth from the Poetry
Society of America that he was pleased to learn that Ruth liked his review of
her book; he added:

I'm sorry that present conditions in the publishing world hampered Decker
in the printing and paper, but a fine book is fine regardless of trappings. And
he's [James Decker] quite a hero in keeping his progressive imprint alive.[19]

Ruth kept working in New York and began writing juvenile fiction without success.[20] Not until 1973 did she learn the story of the Decker Press, and
then from another publisher, Carroll Coleman,[21] who operated the Prairie
Press in Iowa City and published her third volume of poetry, *The Shadow on
the Hour* (1956). He wrote:

I visited the Decker Press three times. The first time was about 1940. I do
not believe he [James Decker] had been going very long then. . . . After an
inquiry at the one filling station in Prairie City, I was immediately directed to
a small room at the back of his grandfather's drug store. . . . At the back of

the store . . . was the Decker Press with James Decker, his sister Dorothy, a clamshell press, and a few cases of type. I believe he had only enough type to set three or four pages at a time, so he would have to print one page at a time. . . . James did most of the typesetting, Dorothy did a good deal of the presswork, and between them I guess they managed the orders, mailings, etc. . . . The binding was done elsewhere, and on all of the books I have, very badly indeed.

My second visit . . . was about 1944 or 1945. . . . and they had the whole building. . . .

On my third visit in the fall of 1949 Decker had gone. He had got badly in debt, was unable to get paper and other supplies, and a fellow named Irvin Tax, from Chicago who had a manuscript with Decker . . . took over the Press. Decker secured a salaried job with some religious publishing house . . . and Dorothy had stayed on. . . .

Then, a few months later, in April, 1950, Dorothy Decker shot Irvin Tax in the back of the head with a rifle, and then shot herself. This was the "abrupt and tragic end of the Decker Press." . . .[22]

From its opening in 1937 the press published volumes of poetry in printings of a few hundred. Decker never developed a marketing program, but realized that poets were an inexhaustible source of material. He soon began to require subsidies. In financial trouble, Decker sold the press to a local lumber dealer who then sold it to Irvin Tax. Decker had contracted to print a volume of the lumberman's poems but failed to bring it out. The Deckers continued to work for Tax, but in 1949 Decker confessed to embezzlement and left town. Dorothy stayed, as she had fallen in love with Tax, who did not return her affection. In May of 1950 after Tax had avoided Dorothy for some months, she ended it for them both.[23]

Atomic Age

In August of 1945, the Second World War ended in the atomic blasts at Hiroshima and Nagasaki and the world entered the atomic age. Ruth was among the poets who responded.[24]

Millions of displaced and homeless people across the world were starving and dying. How can individuals turn inward and avoid the uncomfortable sight? Ruth wonders. And so in the United States the fear of communism replaced the fear of fascists. The fear of nuclear annihilation grew, and the postwar generation began to practice for their own deaths in school with frequent duck-and-cover drills. Communities began to rehearse civil defense practices, and bomb-proof shelters were built for the highest ranks of government. And what, Ruth asked, of us?

Toward the end of the second decade of the twenty-first century, the old fears of nuclear holocaust are renewed. The important thing, as anthropologist Jules Henry wrote in *Culture against Man* (1963), was then as it is today to keep the flame of fear burning to fuel the consumer economy. Today we see starving millions in other lands, fear the atomic bombs of other nations, and live the future of which Lechlitner wrote in "Lines for the Year's End" (1948). What, we may all ask, of us?

Lines for the Year's End

> *The dropped fruit lies beneath its tree,*
> *And this that was our future, now*
> *Becomes a seed for history.*
> *Sensing the root, the bud, the bough,*
>
> *But not what worm will strip what leaf,*
> *By whose economy shall I*
> *Measure my single grain of grief*
> *When millions starve and die?*
>
> *Possibly there is one who can*
> *Before that fact stand blind and dumb*
> *And probe the private world of man;*
> *Or, subsidized by faith, become*
>
> *The minor prophet of an art*
> *Singing an individual song*
> *Sure of a surplus in the heart*
> *A bathroom lark, off-key but strong,*
>
> *A calm, detached observer of*
> *The "common" man against the rich:*

Cushioned by middle-classic love,
Scratching his very personal itch;

Who, from his heat-conditioned bed
And daily breakfast egg, may
For the "ill-housed, ill-clothed, ill-fed,"
Solution by psychiatry;

Or, shelving Freud with Marx, invite
Faith by salvation redefined:
Words make the man—behold the bright
Semantics-liberated mind

The old year ends. This well may be
Our final day—this winter snow
May fall its last on field and tree.
To bed, to sleep go those who know

Our atom-dwelling God will spare
From all mankind a chosen few
To build a deathless future. Where,
Lacking that grace, go I, go you?[25]

For Ruth and Paul, the answer was California. They were headed west to build their own salvation in Sonoma, California. After that, the balance of the natural and social in her poetry shifts. But first they had to find a place, so Ruth took over childcare duties while Paul went west to look for land.

13

CALIFORNIA

Snow

Early in 1946 Jerre Mangione wrote a chatty letter to Paul and Ruth and asked, "How about you three? What is happening to the California-here-I-come project?"[1] California was in their sights, as it was for the Coreys' neighbor, fellow Iowan Darrell Huff, a New York editor chafing at the bit of the rat race of publishing whose wife urged him to quit and freelance.[2]

Both Paul and Ruth had complained of the cold winters and related health problems.

> We had a 1000 feet of driveway to the Post Road. In the winter it would
> fill up with snow. We would shovel for two or three days to get it clear.
> Then we'd drive down town and get groceries. The next morning it would
> be snowed full again. Or guests would come out from the City for the
> week-end. Saturday night the drive would be snowed full. So my wife and
> I would have to shovel all day Sunday and Sunday night to get the drive
> opened up so that I could get them to the train Monday morning. . . . After

DOI: 10.5876/9781646422081.c013

three successive winters of that kind of weather we decided there must be a better climate.[3]

Charlton Laird, who had rented one of the structures on the Corey place while he was a PhD student, returned from the San Francisco area and praised the school system and the climate of California. Huff was interested in moving west and building a house, so he and Corey decided to work together to select a site in California. With the war still on in the winter of 1944–45, all they could initially do was investigate from a distance by writing to local chambers of commerce and the Weather Bureau in Washington, DC.[4]

Sunkist

After he settled into a job at the University of Nevada, Laird agreed to drive west in the Corey's 1930 pickup so they could pick it up in Reno, not too far from California. Then, on April 21, 1946, Huff and Corey launched "Operation Sunkist."[5] Paul, Darrell, and a third friend drove to California to look for places to build houses and live. Paul's next letter to Ruth was from Escondido, California, about a week later.[6] He praised New Mexico and Arizona, but friends there had warned of bad school systems, so he had ruled them out as places to live. He wrote of the miserable heat of the desert, the bad food, and high prices until they came to the mountains where the air turned cool and the surroundings green. In fifty miles, they passed through some country that looked like Putnam County except for the occasional palm tree, some like Iowa, and some wild and rough. The temperature became cooler and cooler until they reached Escondido, where it was cold.[7] By then, having been overcharged for a meal in Tucson, having eaten a bad breakfast and worse lunch, Paul was expecting to be taken at every opportunity but was surprised to find a reasonably priced motel room and a good cheap supper to boot and walked the streets to see stores with prices lower than those in Peekskill. Because he'd read and heard the area would turn brown by July, he wanted to go farther north.

He next wrote from Los Angeles:

I got here to Maxine's last evening—Oh brother this town is a mess. But Maxine and Louis have a marvelous place high up on the hillside overlooking the town. . . .

Sunny California is grey and dismal this morning much to the consternation of the Californians. I still feel that farther north is the place, but I've
found nothing too discouraging yet.[8]

His next letter was from Santa Cruz to say that he and Darrell would look
over each area and find one they liked and then search for particular places
for them to build. He mentioned the chilly fog that came in sometimes as
well as the valleys and their breezes that did not extend to the surrounding
heights. The next letter described several properties they'd looked at and
the positive and negative points of each and the schools of the area. In the
following letter he wrote that he wished that he had his heavy army boots
as they'd been climbing mountains to look at sites, but none was a good
prospect because of lack of roads. "It's hard to find out any real dope on
climate—everybody thinks every other area is abominable. And everybody
contradicts themselves regularly and constantly."[9] They would continue
looking in the San Francisco area. Everywhere there were orchards.

From Novato he wrote of their trip through San Francisco, "full of strange
things like fish wharves," the fog, and hills so steep they had to drive up
them in second gear and the same on the other side. They were disappointed
they'd not yet found a suitable place but would continue north to Sonoma.
There follow a couple of chatty letters. Then he wrote about the weather
and prices in Santa Cruz and Sonoma. In Sonoma, he found two places with
good road and utility access. He wrote over the next few days that they had
revisited the Santa Cruz to double-check a property in which Darrell was
interested.[10]

On May 20, Paul wrote to Ruth that they'd narrowed the search for land
to Sonoma because the school situation was best there and the climate for
sinus problems better than anywhere along the coast. He concluded, "We've
got to be winding things up pretty soon and starting back. As I figure it, we
must have near $3,745.47 [about $54,000 in 2021] in the bank there," and that
he would like to buy a piece of property that would leave most of it in the
bank.[11] The next day he wrote that he and Darrell had agreed on one piece
of land they wanted to look at again, and that they'd spoken with the school
principal and learned there was no tuition in California.[12] He wrote to Anne
to remember her fifth birthday and to tell her about the possibilities of the
land they were considering; that they'd looked at it three times and it looked

better each time, that there were good places for children to play and places for orange and lemon trees.[13]

New Homestead

Paul and Darrell finally agreed on a place, a price, and a division of the land. The 17.83-acre plot cost $3,000, of which the Coreys' share was $1,500 [about $21,600 in 2021]. He and Darrell would bring the title back with them so Ruth and Fran, Darrell's wife, could sign it to put the land in the names of all four people.[14] The Coreys would get 5 to 7 acres, some would be held in common, and the Huffs would get the balance. Corey gave Huff first choice in the division, and both were pleased with their choices. Corey said they needed a survey, but the parcel he'd stepped off was as much as he wanted to handle. He went on to discuss a well. They would leave on May 29 and stop in Cheyenne, Wyoming, to check general delivery for mail and about three days later would be in Cedar Rapids, Iowa, to visit Buel Beems, one of Corey's advisers who had critiqued his drafts of novels and stories. He still had $100 [about $1,440 in 2019] of the $240 he'd started with, so he figured he was doing okay.

At the end of May, Paul wrote from Wendover, Nevada, that they'd met with Charlton Laird and his second wife Helene, the couple who had moved to the Cold Spring land with the Coreys.

"Nevada is full of gambling joints, but I showed character. I put a nickel in a slot machine tonight after dinner, won five and walked away from the machine."[15]

Preparations

Early in 1948, writer David DeJong, who'd been a frequent visitor at Cold Spring, wrote to Ruth and Paul from Providence Rhode Island:

> Astonished, astounded etc. to hear from you out of California, and yes, at the moment wish the hell we were there, too. . . . Have been having the most execrable weather this winter imaginable, snow every other day, and no facilities to move it in this politics ridden town. . . . So we're tired of it, tho understand in California, I guess just about where you are too, unfortunately

you're having a real he-man drought. . . . Just goes to show, I suppose, that you can't have everything that's good. We seem to be having very little of it.[16]

He went on about his writing and the publishing world suggesting "we're sorta riding for a fall. . . ."

Happy to hear you have a roof over your head, and are now husbanding in constant greenery, picking narcissus on Thanksgiving, which seems like a hideous distortion of seasons, but otherwise definitely has its points. I was never one for traditions anyway and would just as soon see June in January, all things being equal in their directions.[17]

Paul was busy organizing and sending the manuscripts for his Mantz trilogy, *Buy an Acre*, and *Five Acre Hill* to the University of Iowa Library. He added, "The brochure you sent me of the new Iowa Library sounds like it's going to be a fine place, and I'm mighty proud to be housed in it."[18] In mid-July the Huffs drove to the land Darrell and Paul had purchased. They towed a mobile home and lived in it while they built a house and wrote to Paul and Ruth of their progress.[19]

Westward the Huffs

Fran wrote that the hardest part of the trip was the mountains of New York. They bought food along the way and cooked it in the trailer, talked with other "trailerites" along the way to compare notes, and reported good roadside parks in Ohio and Indiana. Illinois and Iowa were green with the healthiest and heaviest corn that Fran had ever seen. Their two daughters were traveling well, and by mid-July they had arrived at Fran's parents' house in Cedar Rapids, Iowa, close to Iowa City, where they planned to stay a week to get some mechanical work done on the car.[20] The rear end was sagging. They went to movies and to the Kolache festival in Cedar Rapids, a celebration of Czech ethnicity.

News of the death of a mutual friend moved Fran to write that it was "all the more reason for making the most of each day and not waiting around to retire at 65." She reported that the stores were full of all kinds of food that had been in short supply in the East.[21] Darrell wrote on the eve of their departure to the west that the well and pump on the California land had been completed.[22] Fran wrote, "Most of our time [in Iowa] was occupied by people dropping in at my folks' house every evening—people I hadn't seen for 20 years and that I could well do without seeing for another 20. What misery!"[23]

Fran sent postcards along the way to update the Coreys.[24] A week later, Darrell wrote from Sonoma that he had the trailer parked, the weather was good, and he was searching for tools and materials.[25] A major problem was scarcity of any kind of building material, so if they saw something they needed, they had to buy it and take it or it would not be there later. The weather continued to be dry, but, Fran noted prophetically, everything was as dry as tinder. The people were friendly, and the Huffs were living on the royalties of two previously published books because Darrell wasn't getting much new writing done. "We are happy . . . in spite of our itching [from poison oak] and we certainly aren't looking any thinner. This climate makes one sleep and eat very well. It is too bad that one must do other things too."[26] Toward the end of August, she wrote that no one spoke of weather because it was always good.

Wreckers in San Rafael were tearing down a mansion and selling the materials, which were going fast as many people were looking for building materials. The Huffs' search for and acquisition of building materials continued, but the food was plentiful and delicious.[27] They were cooking outdoors on a charcoal grill and bathing under the sky as they finished collecting their building and plumbing materials. It proved impossible to find a four-inch soil pipe for the sewage, and Fran appealed to the Coreys to send one from New York if they could find it. The Coreys were able to send that item, and the Huffs had acquired an electric stove that would "do everything except buy the groceries."[28]

About the same time, Elizabeth Clarkson Zwart wrote from the *Des Moines Register* that she had heard about the Iowa colony in Sonoma, where everyone was "so rugged that you build houses with one hand and write articles and books with the other."[29]

The power company had to dynamite holes in the bedrock to set poles to bring electricity to their place, and the bulldozer couldn't completely level the building site because of extensive rock, but the Huffs had recruited some help to pour the slab for the cabin.[30] Darrell wrote asking for more sewage pipe and detailing the beginning of construction of the cabin. Two days later he wrote that the electricity was connected, that he'd received the pipe from Paul, and workers were busy on the slab.[31]

"I feel fine," wrote the pregnant Fran, "except for a few back-aches caused by house building rather than infant production. . . . Now that I am giving my all to house-building I appreciate more what you did in building your place."[32] In five months time they could move in. It took eleven more years and two children to complete the house. Darrell divided his time between writing and building.[33] In April 1947, Paul, Ruth, and Anne arrived.[34]

Paul and Ruth Follow

Paul Corey picks up the story:

> We sold our nine-room stone house, loaded up our daughter and two yellow cats and headed west. Here in the Valley of the Moon, across from the Jack London Ranch we began a repetition of the pattern of our lives. We lived six weeks in a rented trailer, while I built a cabin. We lived six months in the cabin until I had enough house built to move into. Then of course the money ran out. My wife became a substitute teacher. I modeled for how-to pictures. Then I became a furniture designer. But it took several years to re-orient ourselves and get the house finished. Our standard of living had improved. This house was all electric. When the power goes off, as it frequently does, you can't even go to the bathroom. . . .
>
> Of course, after several years I had to start building another house. This one was higher on the ridge above our first California house. It has a 300° view. The other 60° we ignored. We sold the other house and moved into this new one with our three cats. We haven't had a yellow cat for eight years now.[35]

Characteristically, Paul published the list of expenses for each phase of the building. The total for the initial cabin came to $386.33, or about $5,300 in 2019. For the complete house of 1,296 square feet of the final, larger house, the total cost of materials was $4,432.84, or just over $62,500 in 2019, all out of their income as writers. He concluded:

> We have proven that low-income people can build a low-cost, high-quality home out of income on the time they have free from their regular working time, if budgeting and planning are handled carefully.
>
> Anne was able to move into her own room on the eve of her seventh birthday.[36]

Ailing Ruth

As they settled into California, Ruth found work as a substitute school teacher and continued writing and submitting poetry, some of which was published. She also prepared and began to submit a collection of her poems. The couple participated in the local community: schools, a play reading group, and conservation efforts. After a decade and a half, Ruth's health began to deteriorate, at first slowly and then more markedly. In April of 1964 at the age of sixty-three Ruth wrote to Malcolm Cowley, the critic and literary historian who was then at Princeton University:

> Currently, after over a year of illness first with bursitis, then with a fractured vertebra which landed me for a while in hospital, I'm just recovering strength enough to begin assembling selected and new poems for a possible fourth book collection; also working on two rather long descriptive-narrative poems.[37]

But the chief reason she wrote to Cowley was to ask his advice on publishing a 5,000-word reminiscence of Robert Frost when he was poet-in-residence at the University of Michigan. She made it clear that during her senior year, in 1922–1923, she got no inspiration or help from him whatsoever and that she found him to be "a completely reactionary, narrow-minded jealous and self-centered person behind his outwardly genial and folksy mask."[38] Ruth wrote the piece in response when her friend, Detroit freelance writer Dorothy Tyler, asked for Ruth's personal impressions of Frost at the University of Michigan when she was there and her estimate of the inspiration Frost passed on to her and the rest of the students. While Tyler did not appreciate Ruth's response, she did send it to Frost's official biographer, Lawrance Thompson of Princeton. He found it useful and wrote to Ruth for more. Thus started a correspondence that lasted until Thompson's untimely death in 1973.[39]

Visit with Daughter Anne

Ruth's letters began to take on a disparaging tone toward other poets that only increased in intensity with time. Toward the end of April 1964, Ruth wrote to radio producer Norman Corwin in Sherman Oaks, California:

Do please forgive me for not having answered your very fine letter to
me dated September 30th. I did thoroughly enjoy everything you had to
say—particularly your paragraph on dear old peace-lover Norman Cousins
and, may I add, his toady poetry editor, John Ciardi, whom I suspect had
something to do with turning down your poems. Fact is, John C can't really
bear any competition in the satire field of poems—he thinks he has a corner
on that market in S[aturday] R[eview]'s "limited" space for poetry.

Why I didn't write before is that I've had a hellva time physically since
November. The day after Kennedy got it in Texas [November 22, 1963], I
fractured a vertebra low in my back, and was scarcely able to navigate for
some weeks—finally they decided to put me in the local hospital for Xrays etc
and found out that I'd inherited a bit more than poor Kennedy's "backache,"
which I'd rather thought. Since then, until just a couple weeks ago, I've been
in sort of a straight-jacket [*sic*] deal—probably a good thing for a poet like me.
Now at last I can walk again, do my cooking and dishwashing, tho no heavy
housework.

Best news for me is that the Doc has given me the go-ahead signal to fly
down to Riverside and visit a couple weeks with my daughter Anne and her
husband [George]. I'm hoping to go down there early next week; some time in
May, Paul will drive down and pick me up for the return trip—by that time my
back should be strong enough to take a long car trip, which it isn't right now.[40]

She concluded with an invitation to visit her in Riverside to discuss "the sad
state of poetry, the world and all that."[41] He answered with a plea of a full
schedule but invited her to visit him so he could "meet you at long last, and
buy you a beer to weep into about our dear dead art."[42]

In May 1964, Ruth flew to Riverside to visit Anne. Addressing her as
"Darling," Paul wrote that *The Christian Century*, with her poem "Moment
Electric," had arrived in the same mail delivery with her letter to him. "I'm
glad that one of the return envelopes carried something besides the usual
rejection," wrote Paul.[43] Ruth also received a chatty letter from San Francisco
poet Kenneth Rexroth, who said that he had looked for a house in Sonoma
and continued, "When you get restless and fretful for travel, just consider
that today it is utterly impossible to improve on where you are."[44]

Ruth wrote to Paul and complained about the train engineer who sounded
the whistle all the way through town at midnight.[45] Paul wrote, again

addressing Ruth as "Darling," and concluded, "I have loved you honestly and I still love you honestly. I love without benefit of conventions. I cannot love by conventional fiat."[46]

Ruth answered:

Hang conventions. If I had ever believed for one minute that you did not love me honestly, the choo-choos that honk from midnight on would have had a passenger long ago. I love you—perverse old bag that I am—each year more and more.[47]

Anne wrote to urge Paul to drive down to visit them because she and her husband probably wouldn't be able to visit Sonoma until Christmas. She continued:

This is one reason why we wish you would come down now, which means sometime in the next few weeks and at your leisure, to take mother home.

We are being careful with the old lady so that she won't have a set-back or get involved in a quarrel. She is sleeping well, mainly, I think, because we, or I, force her to stay awake after dinner until the time she goes to bed. In fact, she went to bed just before eleven last night and didn't have to get up until six, and then I had to wake her up this afternoon after she'd slept about two and a half hours. . . .

. . . I've also taken Mother out for drives with me. . . . Yesterday we drove around in the car for over two hours, and, so far as I could tell, she was comfortable and could have sat longer.

Mother insists that she will be able to travel back by car. She talks almost incessantly about you and fusses and fusses about you not feeling well and looking yellow. Of course, your last two letters to us have sounded rather depressed and you did say yourself that you weren't feeling well. If you really don't feel up to making the trip, I guess it can't be helped, but it really isn't so bad to take it in two days. . . .

. . . Mother has planned a visit to her friend Norman Corwin in Sherman Oaks, and George has to go along because he can't stand me on the freeway. I guess he remembers having a good time with Upton Sinclair, too.[48]

Anne went on to discuss possibilities for the design of a cabinet that she hoped Paul would build for her to hold stereo components. In the same letter, Anne told Paul:

Mother doesn't seem very happy about going home. She really is an independent old witch! Every few days she brags about only having to take one aspirin or maybe none during the day. Apparently her troubles are getting better. I haven't been able to test her on any long car trips. I gave her hell one day for sissing when I put the brakes on a bit fast. She, of course, <u>has</u> to sit with her knees crossed instead of being prepared with both legs or feet on the floor. She claims her sissing is only because she remembers how much bumps <u>used</u> to hurt.[49]

Robert Frost

In July 1964 Thompson responded to Ruth's answer to his query about Frost:

"He Who Rides a Tiger" is a very powerful and moving poem. But even if Frost had been given a chance to see how he understood it, I don't think he would have accepted it as applying to himself. Sure, he would have said, the life of the literary artist is a brutal tiger-ride, and the art is the brute that makes the man-artist brutal; but he would never admit that his art had been that brutal, or that he had ever been a brute. But he was. <u>He would never quite face up to himself</u>, on matters like that. Maybe he had to cheat to keep his self-respect. But cheat he did, whether he had to or not.[50]

Ruth's poem "He Who Rides a Tiger" appeared in her 1973 collection, *A Changing Season*.

He Who Rides a Tiger
  Barred night bleeds darkly down the moon-chill pelt.
  The Tiger-mounted clings to the slippery flesh
  And dares not leave: he takes both pride and guilt
  On the known trail where fact and dream enmesh.

  His treacherous Beast, earth-pugged and heaven-winged, snarls
  From a mouth once dewy with an innocent seeking
  But now greed-reddened. Legs astraddle, galls
  Under his thighs, knee-chafed, hand-blistered, reeking

  Of animal sweat, he tastes repeated fear;
  Sleeps (or dreams that he does) for a brief moment's

Respite from hunger, eyelids closed, to hear
Music like muted bells, like an aloneness
After shared need, or an Aladdin-land
Mist-contoured from an alien dimension
Whose genie leaps from the mind's morning hand
(Is the Beast round with cub?) with new compulsion.

Those who would love him, scream from the rending jaws;
He sees their warm flesh by the Beast devoured;
Frozen, he watches the sharpened scimitar-claws
Slashing the son, abandoned, whom he sired.

He rides because he must. He who has eaten
With his own jealous Tiger, he who with love and hate
Grips his demanding Beast, he who has killed . . . forsaken,
Holds in his heart his self-engendered fate:

Chained to the Beast-spored jungle-carrousel
That circles an ego-centered rod, his spinning
Voice echoes upward from his wound-deep hell
Until a slowing-slowing!—leaves him grinning

And he lets go, falling . . . and falls, at last
By the tired Tiger that he rode, befriended.
Time writes his legend after his day has passed.
But the punishing ride is ended.[51]

Ruth had not yet let go of her own tiger. She wrote to Thompson about a
forest fire in November 1964:

> Paul and I are just digging out from under soot and ash after a recent
> harrowing experience with a huge forest fire. Our new house—the last of
> our home-made homes—was directly in the path of a fast-moving wall of
> fire. We, together with a few grabbed up Ms and books, and three howling
> Siamese, were evacuated midnight of the 22nd. All night, on the deck of a
> friend's house across the Valley, we sat and watched that stretch of 10,000
> acres of timber, ranch houses, out-buildings (to say nothing of wild life) seem
> to dissolve in a 30-foot wall of flame. We didn't even dare to believe that we'd

have a home left to go back to. But we'd built this house high on rocks, with quarry rock half way up, and (take notice, Edna Millay) this fact, plus the hard work of two Forestry-service pumper trucks stationed at our place all night, somehow saved us. (Many other homes however, by a freak change in the wind, were not so lucky.)

But everywhere we look, except down across the vineyards toward Jack London's eucalyptus grove, is a surrealist scene in black of soot-covered ground, white-piled ash of brush, tall black skeletons of trees. Other trees, completely singed by heat are brown-leaved. But those sturdy live oaks, though. . . .[52]

Ruth told Thompson in the same letter that she was occupied compiling her poems and trying to find a publisher: "I've been damned busy getting together a collection of selected and new poems for possible book publication. I'm not very optimistic but damnit, I can try." She added:

I haven't had much luck with my new collection of poems since I started sending the Ms around a couple of years ago. It first went to Rutgers University Press. The editors there sent the Ms to Archie MacLeish for a reading, and being an old dear, he responded gallantly. They seemed much impressed. Then, after holding the ms for about 8 months (!) they wrote to me that, since they already had too many commitments on poetry, with deep regret they were returning etc. Hm . . . Since then several of the larger publishers have returned the Ms with similar "regrets."[53]

Alan Swallow from a press in Denver asked to see a recent collection of Lechlitner's poems and indicated that he would report in January or February. But before the end of the year he died, and Lechlitner tried to get her manuscript back. But no one in Denver knew what to do. She continued as doggedly as Paul was accustomed to doing.

It was to Thompson that Ruth wrote about her major operation in August 1966 with the anesthetic sodium pentothal:

From 8 AM to 6 PM (though only a second in my recollection) I was way out—in the most blissful state of complete non-existence—no bad, no good, no pain, no joy, no space, no time. If death is like this, then death is the most wonderful thing having lived can give one. . . . The only thing was that I had to come back. Back to pain, to worry, to trying to catch up on work, to living in increasingly bad Birchite-filled California. Am I the complete pessimist? Nope. I love these hills, these vineyards, the wonder of the spring flowers that are now bursting out all at once after the winter rains. And relish our good friends and good books . . . and any number of things as you well know. But the very best thing I've ever known is that complete oblivion.[54]

Ruth suffered more than twenty years before she died.

A Writer's Life

In 1968 Paul summed up his career in a letter to a local Iowa librarian:

Every writer, I suppose lives several lives either consecutively or successively. My lives have come in succession and I guess I'm on the fourth at the moment. In the '30s I wrote mainly short stories. In the '40s most of my adult novels and young-adult novels were written and published. At the end of the '40s came the MacCarthy [sic] Era (Joe, I mean), and book publishing went sour. I thought for a time that I might have to change the title of THE RED TRACTOR to THE GREEN TRACTOR. In the '50s I changed to another form of writing—how-to-do-it. I designed a lot of furniture, photographed it, and told how-to build the stuff for the How-to magazines. The witch hunters didn't find me in that disguise. Besides there was money in it and we had a daughter to bring up and get through college.

However, I've always preferred fiction writing and for amusement during the '50s I started writing Science Fiction short stories. A writer can say something in SF without being labeled. But no magazine would listen to me in the US. In the beginning '60s the how-to field went sour. I turned more and more of my attention to SF writing. Still no luck in this country. But an English magazine took one of my stories. In May of this year [1968] a collection called NEW WRITING IN SCIENCE FICTION is using one of my stories. . . .

Shortly after this started happening I got an offer from Edward Hale, Ltd. For my SF novel THE PLANET OF THE BLIND, which I presume will come out sometime this year.

Of course, the how-to field isn't completely washed up. Last June my American publisher, Thomas Y. Crowell brought out a non-fiction book of mine HOLIDAY HOMES which tells people how-to get a "second home."[55]

Rejection and Publication

Now Ruth was getting a taste of the decade-long process of trying to find a publisher. In 1971 she wrote to Thompson:

> I've been dreadfully busy revising many poems, trying to get a new collection together for them; trying to find a small but respectable publisher. During the past four years have sounded out some 30 of the Establishment publishers, been turned down by them all. Well, at age 70, I'm still writing, still trying.[56]

The book, *A Changing Season*, came out in 1973 from Boston's Branden Press. From that book:

Inside and Outside
  Wired for more than I need, this house
  cages me with a network tension
  of nerves twisting into
  a box like a metal brain.
  The Appliances—handy for heating & cleaning
  & freezing & cooking & whipping & drying—
  hum spin gurgle belch grumble spray.
  What men have designed to make easier all
  the routine chores of living, becomes
  a mechanical, many-armed Siva
  for women to worship—or perish.

  The snap of a switch that gives me
  sight after sunset, keeps me
  sleepless, ear-chained, to listen: outside,
  cables thick with Today
  swing over and under air, earth and water
  pulling from cold night
  a hot electric speech.

  There are no more spaces to turn
  around in with comfort, to let me
  breathe when my ribs are parentheses
  closing too much:
  wars, warring ideas, slogans; competing TV

commercials. And after the late news, Sleepeaze
nightmares about violence, hate . . .

What were once small things
have stayed to grow like ships
built in bottles: ironies of illusion
that have no exit.

I can't join the nostalgic who dream
of rooms cordless, with candle or lamplight,
(did they ever I wonder clean
even one of those smoke-sooty chimneys?)
—of the stew kettle craned over woodflame
(well, at least no broiler to scrub),
—carved wood picturesque on a spindle: ah,
that oldentime rocking chair leisure!

Okay. I can split—even freak out
in my turquoise Hawk on the highways,
adding more car stench to the already
fart-stinking air. And my smogged eyes can still
see those ditch-held cartons of garbage
fought over by rodents
—and by the unwanted, half starved
cats that have also been thrown
from Humanity's cars.

A new layer of beer cans, foil, broken glass
replace what school kids on Earth Day
picked up,—along with
a new word: Ecology.
But what I like best are those countryside
Kleenex bushes—twig-caught scraps
of rose, white and blue like the faded
and tattered remains of Old Glory . . .

The machines inside, the destroyers outside,
Blood brothers (or kissing cousins)—each
dependent on the other.

What shall I do with the time saved
by a faucet that pours
treated and heated water
into coffee that's instant?
—I could order a new cord for
the 7-speed Marvel blender,
—or scan the mag digest of
a book called "America's Greening,"
—or I might let my heart flow freely out
toward simple wholeness,
to find on the untouched edge
of otherness, something complete
in itself: beyond my kitchen door
that slim young locust tree
proud in her flowering
First Communion white.

Inside or outside,
the fraction of time that's left me
tells me survival can still be
a matter of choice.

And I choose.[57]

Lawrance Thompson

Lawrance Thompson continued his biographical work on Robert Frost and his correspondence with Ruth, writing in 1971:

> The third volume will be interesting enough to me, whenever I find the energy to get started on it. In some ways, however, it will be the most difficult-delicate of the three—for reasons you may or may not know: Frost fell in love with another man's wife, just three months after the memorial service for Elinor Frost in Amherst. (Hope that fact doesn't spoil the last chapter of RF-II for you.) For several years thereafter, Frost did his best to get the woman to arrange a divorce so she could marry him. (I was in on it, and urged her <u>not</u> to get a divorce!) How do I handle these delicacies? Well, I've already had a little practice, as you will see if you look carefully at

pages 465–484 of SELECTED LETTERS OF ROBERT FROST, and page 497–8, a letter written F finally became convinced that she would never get the divorce; would never marry him. The man and wife are still alive, and the question of how I'll handle the story (which really continues throughout volume III) will depend on how much they will permit me to tell. It's a question of taste and I'm forced to tell some lies, or at least forced to skirt the truth somewhat vaguely. Once again, the game will be to write for readers with your gift of reading above-below-behind-between the lines.[58]

Another Visit to Anne

Tillie Olsen opens her *Tell Me a Riddle* with lines that could equally well come from the lives of Paul Corey and Ruth Lechlitner at this time:

> For forty-seven years they had been married. How deep back the stubborn, gnarled roots of the quarrel reached, no one could say—but only now, when tending to the needs of others no longer shackled them together, the roots swelled up visible, split the earth between them, and the tearing shook even to the children, long since grown.[59]

In 1975 Paul wrote to Ruth, who was visiting Anne, now in Hawaii, "You're right, this February 1st is our first anniversary apart in 47 years, which represents about a 40 year delayed reaction. . . . We have put up with each other for a mighty long time."[60] He went on to catalogue their assets and suggest an amicable divorce.

That never happened though concern for the delicacy of the details dictates their omission. Paul and Ruth would stay together for the next fourteen years. But it is clear that whatever satisfactions they may have derived from their writing, building, and daily lives, from neither perspective was their intimate life a success.[61] Paul spent more and more of his time tending to the increasingly ailing Ruth. Both of their memories were failing, and each relied on the other to help fill in the increasing blanks.

Spiral

In 1984 literary historian Douglas Wixson visited Paul and Ruth at their house in Sonoma Valley. He wrote:

Ruth has suffered several strokes and can no longer manage without Paul's help. Her memory is diminished as a result of the strokes, yet she helps Paul recall names of literary people they have known. Their marriage was from the start a partnership of two creative people and remained so throughout their lives.[62]

That it was, but after the intimacy predicament the partnership became more unbalanced and continued to spiral downward to the end, as revealed in his letters to close friends.

January 1988:
Ruth seems to be slipping deeper into senile dementia, which makes it harder for me to get anything done. However, this morning she's fairly quiet so far, reading a book. That occupation can keep her quiet for quite a while, even though she doesn't remember what she reads from page to page. Sometimes she worries a lot over nothing, and I find it harder and harder to get her to go out. She can get around all right with her walker but she keeps saying that she can't make it to the car or get into the car when she gets there. I have to bully her into doing things like that. She immediately forgets my bullying but I don't and it's hard to live with.[63]

March 1988:
This existence is not a happy one. My daily schedule goes like this: up at 6:30 a.m., open bedroom door and door into livingroom to let cats come through to the bathroom, then open bathroom door to patio just wide enough to let cats out, meanwhile I've put heat on in bedroom and bathroom and am sitting on john. Ruth asks what day it is. I wash my face in cold water, dress, put her robe on her wheelchair and drops for glaucoma in her eyes, and tell her to take her time getting up. Then I go out to the livingroom, turn up the electric heat, pull on the TV, start the fire in the fireplace stove, and put coffee water on to heat, then feed cats, and make breakfast. Breakfast is usually the same: mixed fruit and cottage cheese fruit for R, plus her morning pills, and orange juice. Mine is bagel half, peaches and coffee. Meanwhile the TV is giving the news. By the time the coffee water boils R is clamoring for me to wheel her out to breakfast. Just as she is capable of putting on her robe, sitting on the can, and wheeling herself to the bedroom door, she is also capable of opening that door and wheeling herself out to breakfast, but she thinks she can't and unless I become furious and yell at her, she won't do

it. So I usually go the bedroom door and wheel her out because I've got to keep from being forced into a temper so early in the morning. She pretends she can't see, but I usually ignore that. After breakfast while I drink coffee and watch the news she reminds me that she can't see which is only partially true and is her blue slacks and red blouse laid out on the bed. I've got tired of protesting that she is not wearing them that day so I say they are. I used to explain that her day clothes would be waiting for her when she was ready for them. She can bathe herself and dress herself if I order [her] to in a loud enough voice. But she can't remember much beyond 15 minutes. If I yell at her to shut up and settle down, she may leave me free to do some work that has to be done. An hour before lunch she starts pestering about getting to the table. After lunch she has to be put to bed for two hours and in all this I've got to cook, dream up menus, take care of business, do the things that have to be done about the ranch, and correspondence, cook, wash up, look after myself, and recently, I've been trying to write an article—Then there's shopping, chauffeuring, going to the dump, setting recyclable items to the recycle station.

So, if I happen to be a bit slow in replying—allow two months. If you don't hear by then, well, hopefully I may have dropped dead. I hope I don't have a stroke. [He provides phone numbers of neighbors to call.][64]

July 1988:

Shortly I'll be 85 and at the moment relatively in good shape but there may come a time when I won't be. In that event I hope to be able to do something about it [re Hemlock Society]. I don't want to live in a lot of pain and I don't want to "exist" at the expense of someone else's time and energy and life the way Ruth is doing. . . . There is no will that can tell caregivers to "put down" a person when he/she is suffering from senile dementia . . .

I don't know if I mentioned our daughter in letters to you or not. She's 47, has been married and divorced twice, has a son by the first husband, who has just graduated from high school in Ventura where he lives with his father and stepmother. Our daughter Anne lives 20 minutes away and works at the state hospital across the valley from us. She claims to be very busy but doesn't get along with her mother. I think she has the feeling that her mother is simply "using" me and is not as badly off as she pretends. This could be a real thinking or a convenient one to get herself out of the job of helping out. She is now living with a nice guy, who I think she is under the impression that she is rehabilitating. She has had "rehabilitation syndrome" from the beginning. I don't know how she got it. . . .

. . . people say "I'm a saint." "I'm such a wonderful caregiver and so selfless, etc." All a lot of shit. I'm trapped and can't get free short of chewing my leg off. I've considered the options. I can just walk out anytime. But what would I live on? What would happen to my cats whom I truly love? I would never be able to find a place with the view this one has. I could bop R on the head with a fireplace tool. (In some of my rages that has crossed my mind) But that would only get me the loss of my cats, my place and a hell of a lot of stupid publicity. You can probably find a lot of ways out for me in your fertile mind. I could put R in a home but I would still lose my living and cats in the long run.[65]

August 1988:

I had occasion recently to dip into my fourth Iowa novel "Acres of Antaeus" which was published more than 40 years ago. It's all about farmers losing their farms to Agri-business just as they are today. It's just as applicable today . . . no one paid any attention then. However your brothers Ed and Fred have done a splendid job bucking the trend in farming.[66]

October 1988:

Your item about "temper" is interesting. I'm glad that I live out in the boonies with space all around. I'm always afraid that someone will hear me when I "flip" over some stupid thing R does. I must be the equal of any sailor, roustabout, or truckdriver alive.[67]

Paul carried on a long on-again-off-again epistolary relationship with a married woman in Southern California named Lyn. In another letter to her he again revealed his feeling of being trapped by his situation:

I am not a "tower" of strength. I'm an animal caught in a trap. I lash about biting at the jaws of the trap, biting at the parts of my body held, biting at sticks and brush and anything that comes near. I haven't yet chewed enough at myself so that the part held will break free. You have never seen an animal caught in a steel jawed trap. I have. I have caught animals in steel jawed traps. Maybe that's why I'm in my predicament. And like any animal so caught, I can't bring myself to beg for help. And I suspect the only real is the kind I gave the animals I caught in traps—a blow on the head or a bullet through the brain.

I don't ask Anne for help. I don't ask anyone for help. There is an organization to help the care-givers of the brain impaired that came to me and offered
to supply someone to spell me one day a week. At first that one day a week
for the month and no chg. Then it was cut to one day a wk and I paid $8 of
the day's wage. Then it was reduced to two days a month. That will last to
the end of the year. If I get a day off the other weeks I must pay for them
myself—$40 for 7 hrs. For some reason I find it hard to ask for help. . . .

I'm trapped. I let myself be trapped 30 years ago, or perhaps I let myself
be trapped by a lifestyle that is basically a comfortable trap. But it takes two
to tango. When the steel jaws grind into my shanks, I lash out at anything or
anybody. And I'm sorry if in so doing I hurt you.[68]

To his male friend, Mel Yoken, Paul wrote in August 1989:

. . . I'm ready to go any time.

My projects: Ruth in a conv h., reduce pet population, save the mountain
lion, zero population, protect the environment, free education for everyone
who wants it, health protection for everyone from c[radle]-to-g[rave], no
more homeless, no more racism, no more trophy hunters, no more religious
fanaticism, pure air, pure water, to get some money out of the publisher of
my cat books, to get the world to put the good from every philosophy and
religion together and eliminate the bad. And after 86 years, I am tired.[69]

On November 9, 1989, Ruth died after a short stay in a nursing home.

On December 31, 1991, Wixson returned to visit Paul Corey a second time,
this time with his wife Suzanne. Wixson concluded:

The strong simplicity, sensitive firmness, and honesty of "Their Forefathers
were Presidents," Paul's literary legacy are like dormant seeds of life waiting
germination that will fructify the land by which it was nourished.

Evening comes, the house grows dark. Paul is visibly tired. We descend the
mountain road in the gathering dusk to our motel in Sonoma. My last mental
picture of Paul Corey is a small man, slightly bent, standing in his driveway
to bid us goodbye, turning around slowly to enter an empty house.[70]

On December 17, 1992, Paul died.

EPILOGUE

Corn, soybeans, pork. Today these are Iowa's chief exports aside from tractors. Along with people, Paul Corey is among them. Or, in the more contemporary words of the 1938 Federal Writers' Project *Iowa: A Guide to the
Hawkeye State*:

> The younger generation showed a definite reaction from farm life, and many
> of those who left Iowa did so to find a more congenial field for artistic expres-
> sion and what they believed could be a richer human life. But memory of
> the prairie soil and skyline was reflected in their work, and the lives of rural
> people furnished the theme for novel, poem and picture.[1]

To write honestly and realistically of his neighbors in the grip of pride,
greed, avarice, and sexual longing, Paul Corey had to leave the place because
no Iowans could admit to such feelings, much less talk about them.[2]

"Corey's antiheroic subject and collective approach to the novel require that,
in the end, the characters fade back into the community," wrote Douglas
Wixson.[3] Like the gospel of work by which he lived, like the family farms

DOI: 10.5876/9781646422081.c014

whose lives and deaths he chronicled, Paul Corey recedes into the fog of time, invisible in today's Iowa with its interstate freeways, microwave towers, 10,000-acre operations, water pollution, corrupted politics, and thoroughly commercialized life following the gospel of wealth.[4]

However, across those wastelands are wisps of hope in a new generation struggling against the same foes Corey described to establish themselves on modest pieces of land in Iowa to return food production to sustainable regenerative methods by which they can live and from which they can feed Iowa, if not the world. The forces Corey described have only gained in strength since his death, but the next generation is always a harbinger of hope.

They may yet return the gospel of work to the soil of Iowa. And when they do, perhaps they will discover Paul Corey and claim him as a prophet. Standing beside him will be Ruth, her poetry describing the world around her in sharp detail. And whatever the travails of old age, they will once again hold hands in partnership.

A NOTE ON SOURCES

The unpublished material I have used is in the University of Iowa Library's Special Collections. There are two collections: Paul Corey and Ruth Lechlitner. I do not include references to material beyond the collection, correspondent, and date. Both collections are well organized and extensively indexed.

The Paul Corey collection is divided into correspondence, subject files, manuscripts, scrapbooks, and addenda. The correspondence is indexed by the last name of the correspondent, the box number, and the folder number of the item. The precise location of any letter can be found via this index. Ruth's papers are similarly organized. Letters regarding various works are often filed with the works themselves rather than correspondence. The University of Iowa Library's Special Collections has prepared an extensive and detailed inventory of all of the material in the two collections, and they are both available online.

NOTES

Introduction

1. Woodard, *American Nations*, 264.
2. Anderson, *One Size Fits None*, 264.
3. Ibid., 264.
4. Ibid., 147, 266.
5. Ibid., 150.
6. Schlosser, *Fast Food Nation*.
7. Oliver, "Ecology and Cultural Continuity."
8. Ewers, "Review of *Ecology and Cultural Continuity*."
9. Behar, *Translated Woman*; Frank, "Ruth Behar's Biography in the Shadow."
10. One excellent treatment of biography in anthropology with reference to philosophy, literature, theology, ethics, ethnographic fieldwork, and history as well as a detailed list of examples is Langness and Frank, *Lives*. Other treatments include Helm, "Preface"; Gruber, "In Search of Experience"; Blackman, "The Individual and Beyond"; Crapanzano, "Life-Histories"; Dunaway, "Oral Biography," 259–262, 265; Mintz, "The Anthropological Interview"; Glazier, "Paul Radin."

11. See, e.g., Lowie, *Robert H. Lowie*; for the New Guinea version, see Banner, *Intertwined Lives*, 319.

12. de Pina-Cabral, "History of Anthropology," 26.

13. Fast, *Being Red*, 31.

14. Salzman, *Understanding Culture*. He is discussing Evans-Pritchard, *The Sanusi of Cyrenaica*.

15. The reference is to the Farm Holiday Association that advocated that farmers keep their products off the market until they received what they considered fair prices. It took its name from Roosevelt's bank holidays. The speaker would have been its founder, Milo Reno, who resigned as president of the Iowa Farmers Union to take on the responsibility and was known as a rousing orator. See Federal Writers' Project, *Iowa*, for a more detailed description of the movement and its leaders.

16. Corey, *Acres of Antaeus*, 275–278.

17. Doukas, *Worked Over*; Fones-Wolf, *Selling Free Enterprise*.

18. Corey, undated typescript in response to letters from Helen Louise Johnstone, Magazine Liaison Section, Office of International Information & Cultural Affairs, US Office of War Information, February 26, 1946, and March 7, 1946. This typescript is titled "The Mid Westerner" and is filed with correspondence about the *Esprit* story published as "L'homme du Middle West," Paul Corey Papers.

19. Corey, "L'homme du Middle West," 566.

20. Mills, "Politics, Power, and People"; Sawchuk, "The Cultural Apparatus."

21. Corey, *Acres of Antaeus*, 22.

22. Ibid., 38.

23. Ibid., 336.

24. Ibid., 386.

25. Ibid., 387–388.

Chapter 1: America's Disappeared

1. For a more complete list, including poets, see Nelson, *Repression and Recovery*, 166.

2. Wixson, *Worker-Writer in America*; Langer, *Josephine Herbst*; Strong and Keyssar, *Right in Her Soul*; Grunfeld, *On Her Own*.

3. Reid, *Tillie Olsen*; Rosenfelt, "Introduction"; Coiner, *Better Red*.

4. Wald, *Exiles from a Future Time*; Wald, *Trinity of Passion*; Wald, *American Night*; Saxton, *The Great Midland*.

5. Pells, *Radical Visions*, 331–334, 347.

6. There is a large literature regarding this. One recent summary is Hochschild, *Spain in Our Hearts*.

7. Folsom, *Days of Anger*, 116.

8. Ibid.

9. Murray, *Red Scare*; Hicks, *Where We Came Out*, 121.

10. Schrecker, *The Age of McCarthyism*; Hicks, *Where We Came Out*, chaps. 8 and 9; Eisinger, *Fiction of the Forties*, 8.

11. Hicks, *Where We Came Out*.

12. Berger, *Dollar Harvest*.

13. Field, "The Price of Dissent"; Field, *Harvest of Dissent*.

14. See, e.g., Fried, *The Most Dangerous Man*, 115.

15. Hicks, *Where We Came Out*, 121.

16. Kutler, *The American Inquisition*; Robbin, *Alien Ink*; Mitcgang, *Dangerous Dossiers*; Belfrage, *The American Inquisition*.

17. Durrenberger, "Malinowski Award Lecture."

18. Wilson, "Introduction," xviii.

19. Kessler-Harris and Lauter, "Introduction," xiii.

20. Hicks, *Where We Came Out*, 59.

21. Geismar, "Naturalism Yesterday and Today," 5.

22. Hicks, *Part of the Truth*; Susman, *Culture as History*; Pells, *Radical Visions*, 365; Fast, *Being Red*, 52.

23. Scott, *Seeing Like a State*, 196–197.

24. Hicks, *John Reed*.

25. Ibid.; Langer, *Josephine Herbst*.

26. Hicks, *The Great Tradition*, 300–305.

27. Pells, *Radical Visions*, 161–163.

28. Hicks, *The Great Tradition*, 303.

29. Ibid., 299.

30. Ibid., 262.

31. Cowley, *Think Back on Us*, 34.

32. Ibid.

33. Ibid., 35; Pells, *Radical Visions*, 88–95.

34. Kazin, *Starting Out in the Thirties*.

35. Folsom, *Days of Anger*.

36. Wald, *Exiles from a Future Time*, 235. Literary historian Wald suggests from a later perspective that "Lechlitner and Corey saw themselves as revolutionaries who believed in ecology and a back-to-the-land ethos as key elements of the Communist project."

37. Folsom, *Days of Anger*, 301.

38. Ibid., 340–341.

39. Kessler-Harris and Lauter, "Introduction," ix.

40. Pells, *Radical Visions*, 168–169.

41. Ibid., 301.

42. Maas to Lechtlitner, December 29, 1934; Maas to Lechlitner, January 6, 1935; Maas to Lechlitner, November 24, 1935, Ruth Lechlitner Papers.

43. Rahv to Lechlitner, March 25, 1935, Ruth Lechlitner Papers.

44. Rosenfeld to Lechlitner, December 19, 1935, Ruth Lechlitner Papers.

45. Blake to Lechlitner, August 31, 1936, Ruth Lechlitner Papers.

46. Lechlitner, ". . . anti-war and anti-fascism . . . ," 79, 80; Corey, "Lurching toward Liberalism." As Alan Wald points out, such recollections were shaped by the political pressures of the times and are not reliable historical guides, but I develop other evidence in chapter 9.

47. Wald, *Exiles from a Future Time*, 1–2.

48. Folsom, *Days of Anger*, 112; Fast, *Being Red*, 3.

49. Quoted in Folsom, *Days of Anger*, 157. Fast, a communist writer, is best known for his 1951 novel, *Spartacus*, made into a film in 1960 by Stanley Kubrick with a script by blacklisted communist screen writer Dalton Trumbo.

50. Fast, *Being Red*, 286–300.

51. Corey, "Lurching toward Liberalism," 67. I have not been able to find corroborating correspondence, and I have found evidence that Corey's account or his memory of the period was blurred.

52. I found no corroborating documentary evidence in Corey's or Lechlitner's papers.

53. Langor, "Afterword," 431.

54. Ibid., 440.

55. Bindas, "Review of Michael Denning," 997.

56. Benson, *John Steinbeck*, 512–513.

57. Langer, *Josephine Herbst*, 290–295; Fast, *Being Red*, 338–340.

58. Ware, *Holding Their Own*, 158.

59. Nekola and Rabinowitz, *Writing Red*, 171.

60. Saxton, "Introduction," xxii.

61. Hicks, *Where We Came Out*, 52.

62. Ibid., 53.

63. Kessler-Harris and Lauter, "Introduction," xiii.

64. Doukas, *Worked Over*; Fones-Wolf, *Selling Free Enterprise*.

65. MacLean, *Democracy in Chains*; Phillips-Fein, *Invisible Hands*.

66. See, e.g., Durrenberger and Doukas, "Class in the USA."

67. Durrenberger and Doukas, "Gospel of Wealth."

68. Wald, *American Night*, 304.

69. Ibid., 316.

70. Eisinger, *Fiction of the Forties*, 4.

71. Pizer, "Contemporary American Naturalism," 270. For an example of natu-ralism in 2019 see Tomar, *A Prayer for Travelers*.

72. Pizer, "Contemporary American Naturalism," 255.

73. Meyer, "Naturalism in American Farm Fiction," 27; Benson, *Wallace Stegner*, 15.

74. Meyer, "Naturalism in American Farm Fiction," 35.

75. Meyer, "Naturalism in American Farm Fiction," 35.

76. Flanagen, "The Middle Western Farm Novel," 125.

77. Meyer, *The Middle Western Farm Novel*, 193.

78. Ibid., 98.

79. Ibid., 175.

80. Denning, *The Cultural Front*, xvi–xvii; Weiner, "Review of *The Cultural Front*." Denning seems to have little respect for chronology. Every instance of Howard Fast's work he mentions—from his work with Voice of America (p. 82) to his his-torical novels (pp. 131, 248)—were from the time before Fast was involved with the Communist Party.

81. Shockley, "The Reception of *The Grapes of Wrath*."

82. Ibid.

83. Geismar, *Writers in Crisis*, 268.

84. Ibid.; Pells, *Radical Visions*, 215.

85. Geismar, *Writers in Crisis*, 240.

86. Denning, *The Cultural Front*, 265.

87. Willis to Corey, March 22, 1984, Paul Corey Papers.

88. Corey to Willis, April 5, 1984.

89. Ibid.

Chapter 2: The Prophecy and its Fulfillment

1. Corey, *Red Tractor*, 16.

2. Ibid., 19.

3. Ibid., 19–20.

4. Ibid., 34.

5. Ibid.

6. Ibid., 35.

7. Ibid., 87–88.

8. The 4-H pledge is as follows: I pledge my Head to clearer thinking, my Heart to greater loyalty, my Hands to larger service, and my Health to better living, for my club, my community, my country, and my world.

9. Corey, *Red Tractor*, 100.

10. See, for instance, Collins, "American Corporatism"; Phillips-Fein, *Invisible Hands*, 31, 61.

11. Schultz, *Agriculture in an Unstable Economy*, 186–203.

12. Ibid., 202.

13. Ibid., 197, 198.

14. Smith to Corey, February 12, 1942, Paul Corey Papers.

15. Ibid.

16. Corey, "Their Forefathers Were Presidents."

17. Corey, "When Farmers Sang," 100.

18. Wixon, "Looking for Paul Corey," 65. New owners of the land built a modern house there.

19. Corey, "L'homme du Middle West."

20. Gerber, "Becoming an Author."

21. Corey to Arthur W. Want, February 5, 1964, Paul Corey Papers.

22. Trachktenberg, *The Incorporation of America*, 20–22.

23. Howard, *Two Billion Acre Farm*.

24. In response to my question to him in 2017.

25. Corey, *The Road Returns*, 291.

26. Miner, *Culture and Agriculture*, 25.

27. Ibid., 26.

28. Ibid.

29. Ibid.

30. Ibid., 27.

31. Ibid., 28.

32. Saxton, *Grand Crossing*, 365.

33. Federal Writers' Project, *Iowa*, 67.

34. Ibid.

35. Culver and Hyde, *American Dreamer*, 116.

36. Miner, *Culture and Agriculture*, 87.

37. Ibid.

38. Ibid.

39. Ibid., 89.

40. Ibid., 90–91.

41. Ibid., 91.

42. Ibid., 93.

43. Wallace, "The Year in Agriculture," 25.

44. Ibid.

45. Miner, *Culture and Agriculture*, 94.

46. Raup, "Corporate Farming in the United States."

47. Federal Writers' Project, *Iowa*, 69–70. The individual contributors remain anonymous. Even attempts by participants and administrators during the period failed to identify them. See Buel Beems to Corey, September 22, 1938, Paul Corey Papers. He and Paul Corey tried unsuccessfully to identify the writer who had mentioned Corey in the guide's section on Iowa literature. Jerre Mangione, who worked with the Federal Writers' Project, suggested it might have been Marshall Best at Viking, the firm that published the guide; Mangione had Paul's name added in the guide. Mangione to Corey, Tuesday and Wednesday [1938?]. But Beems was corresponding with people at the press who said that was not the case, that the writers remained unknown.

48. Durrenberger, "Notes on the Cultural-Historical Background."

49. Berger, *Dollar Harvest*; Block, *The Separation of the Farm Bureau*; Culver and Hyde, *American Dreamer*.

50. Durrenberger, "Notes on the Cultural-Historical Background."

51. Taylor and Taylor, *The Story of Agricultural Economics*, 95.

52. Ibid., 96.

53. Examples include Mighell, *A Study of the Organization and Management of Dairy Farms*; Munger, *The Cost of Producing Milk*; Munger, *Iowa Farm Management Surveys*; Taylor and Hurd, *Farm Organization*; Thomas and Hopkins, *Costs and Utilization of Corn*.

54. Munger, *Iowa Farm Management Surveys*, 376.

55. Taylor and Hurd, quoted in Durrenberger, "The Cultural-Historical Background," 16.

56. Ibid.

57. Ibid.

58. Bowers, *The Country Life Movement*.

59. Busch, *Science and Agricultural Development*.

60. Howard, *Two Billion Acre Farm*, especially chap. 14.

61. Chayanov, *The Theory of Peasant Economy*.

62. Durrenberger and Erem, *Anthropology Unbound*, 189–191; Scott, *Seeing Like a State*, chap. 6.

63. Morrison, *Farmers' Organizations and Movements*.

64. Cochrane, *The Development of American Agriculture*, 378–395; Vogeler, *The Myth of the Family Farm*.

65. Canovan, *Populism*. New York. Harcourt Brace Jovanovich.

66. Field, "The Price of Dissent"; Field, *Harvest of Dissent*.

Chapter 3: Iowa City to New York City

1. Herman Melville was a customs inspector; Corey worked for an encyclopedia publisher.

2. Wixson, *Worker-Writer in America*, 176.

3. Pells, *Radical Visions*, 35.

4. Corey started writing while he was in high school in Atlantic, Iowa, where he worked with Charles Nelson, his lifelong friend, on the *Needle*, a high school newspaper. He recalled those days fondly in a letter of 1945 to his high school English teacher, Mrs. H. O. Henningsen, whom he thanked for not stepping on his exuberance. Corey to Henningsen, October 15, 1945.

5. Corey, "Lurching toward Liberalism," 49.

6. Ottenheimer to Corey, July 18, 1925.

7. Ottenheimer to Corey, November 2, 1925.

8. Corey, "Lurching toward Liberalism."

9. Van Doren, *Contemporary American Novelists*, 149.

10. Ibid., 153–157.

11. Ibid., 161–166.

12. Pells, *Radical Visions*.

13. Kazin, *On Native Grounds*, 207.

14. Langer, *Josephine Herbst*, 88–89.

15. Cowley, *Exile's Return*, 21.

16. Cowley, *Think Back on Us*, 354.

17. Wixson, *Worker-Writer*; Kazin, *On Native Grounds*, 166–171; Reigelman, *The Midland*.

18. Cowley, *Exile's Return*, 48, 223; Pells, *Radical Visions*, 36.

19. Cowley, *Exile's Return*, 69.

20. Ibid.

21. Ibid.

22. Ibid., 72.

23. Ruth Lechlitner, "Biographical" (typescript), 2, Lechlitner Papers.

24. Ibid.

25. John T. Frederick to Lechlitner, May 13, 1935.

26. Corey, "Lurching toward Liberalism," 48.

27. Ibid.

28. Ibid., 50.

29. Ibid., 51.

30. Ibid.

31. Nelson, "My Uncle Paul," 55–56.

32. Unless noted otherwise, ellipses are original.

33. Edwin Arlington Robinson (1869–1935), American Pulitzer Prize–winning poet. The title of Ruth's MA thesis was "Arthurian Story Retold by Edwin Arlington Robinson."

34. Ruth to Jurgen, April 4, 1926, Paul Corey Papers.

35. Lechlitner, "Wife-Thought," 324. Given subsequent events, the allusion to "shining and slim" knives against heaven, juxtaposed to "no child cries out for me," may suggest an abortion. I am indebted to Dimitra Doukas for this suggestion.

36. Ruth to Paul, June 1, 1926, Paul Corey Papers.

37. Corey, "The Pattern," 1, 2, box 12, UIL.

38. Lechlitner, "Biographical," 2.

39. Ruth to Dorothy Hall, February 7, 1929.

40. Cowley, *Exile's Return*, 47.

41. Ibid., 26–47.

42. Ibid., 65.

43. Corey to Lechlitner, May 5, 1930.

44. Corey, "The Pattern," 2; Lechlitner to Anne Spence, 1976(?), 6. Corey recalled that he did offer to marry her. Corey, "The Pattern," 2.

45. Corey, "The Pattern."

46. Ruth to Paul, May 28, 1927.

47. Slang for a Model T Ford or other small, cheap and old car.

48. Dorothy to Ruth, May 26, 1927, Ruth Lechlitner Papers.

49. This is a saying often repeated by Christian Scientists. It is from founder Mary Baker Eddy's *Science and Health with Key to the Scriptures*, 494.

50. Ruth to Paul, May 26, 1927.

51. Paul to Ruth, June 2, 1927.

52. Lechlitner to Corey, Thursday PM, May 1927.

53. Ware, *Holding Their Own*, 63–64; Banner, *Intertwined Lives*.

54. Dorothy to Ruth, August 23, 1923; Dotty to Ruth, September 22, 1973.

55. Dorothy Ivick to Lechlitner, March 9, 1971.

56. Ruth to Paul, June 27, 1927.

57. Paul to Ruth, June 29, 1927.

58. Greenberg Publishing to Corey, September 16, 1927; Simon and Schuster to Corey, November 3, 1927; J. H. Sears and Company to Corey, January 11, 1928; Coward-McCann to Corey, February 1, 1928.

59. Cowley, *Exile's Return*, 79. Cowley includes Paul Corey's name in the list of those he is discussing (314).

60. Paul to Ruth, July 22, 1928.

61. Lechlitner to Spence, 2.

62. Paul to Ruth, July 25, 1928.

63. Paul to Ruth, August 1, 1928, 2.

64. Ibid.

65. This is probably a reference to Baudelaire's 1857 collection of poetry of that title in the plural.

66. Dorothy Hall to Ruth, July 10, 1928.

67. Cowley, *Exile's Return*, 55.

68. Ibid., 79.

69. Ibid., 80.

70. Corey, "Lurching toward Liberalism," 53.

71. Ruth Lechlitner, "Cycling through France" (typescript), 3, Lechlitner Papers.

72. Ibid., 4.

73. Ibid., 2.

74. Ibid., 4–15.

75. Ibid., 16–18.

76. Ibid., 25.

77. Ibid., 30.

78. Corey would use Bandol as the setting for his 1951 story of World War II resistance against the Nazi occupation, "Bridge over the Grand Vallat."

79. Corey, "Lurching toward Liberalism," 53.

80. Ibid.

81. Cowley, *Exile's Return*, 243.

82. Charles Nelson to Bandol, March 29, 1929.

83. Nelson to Paul and Ruth, April 18, 1929.

84. Nelson to Paul and Ruth, May 10, 1929.

85. Corey, *Five Acre Hill*, 9–10.

86. Hicks, *The Great Tradition*, 275.

87. Isidor Schneider to Lechlitner, November 11, 1935; Irita Van Doren to Lechlitner, November 20, 1936.

88. West to Lechlitner, May 4, 1936.

89. Corey, "Lurching toward Liberalism," 55.

90. Cowley, *Exile's Return*, 213.

91. Corey to Virginia Rice, October 6, 1932.

92. Cowley, *Exile's Return*, 298.

Chapter 4: Living on the Land

1. Lechlitner, "Biographical" (typescript), 2.

2. Corey, "Lurching toward Liberalism," 57.

3. There was a large encyclopedia industry during the Depression, and "anyone who was literate and knew a few things could get a job." Moore, "Preface," vi.

4. Corey, "Lurching toward Liberalism," 58–59.

5. Laird, "Bullet Proof Culture." The four authors agreed to use the single name of Laird to avoid confusing the editors.

6. Corey, "Lurching toward Liberalism," 60.

7. Wallace, *Yearbook of Agriculture*, 23.

8. Handwritten note at bottom of Baxter International Economic Research Bureau to Corey, June 30, 1946.

9. Conklin, *Tomorrow a New World*, 8.

10. Howard, *Two Billion Acre Farm*, 209.

11. Ibid., 209.

12. Bowers, *The Country Life Movement*.

13. Danborn, "Romantic Agrarianism," 4.

14. Ellsworth, "Ceres and American Men of Letters," 177.

15. Cowley, *Think Back on Us*, 51, 52.

16. Ibid., 52–55.

17. Seabrook, "Pioneer Spirit."

18. Turton, "80 Years Later."

19. Seabrook, "Pioneer Spirit," 45.

20. Paul Corey, "Prospectus for 'We Bought an Acre'" (typescript), 1963(?), 3, Corey Papers.

21. Ibid.

22. Lechlitner, "Biographical," 3.

23. Paul Corey, "Fugitives from Grayflannalism [*sic*] or We Bought an Acre" (typescript), 1963, 13, Paul Corey Papers.

24. Lechlitner, ". . . anti-war and anti-fascism . . ."; Corey, "Lurching toward Liberalism," 54; Lechlitner, "Biographical."

25. Lechlitner, "Biographical," 1.

26. Ibid., 2.

27. Ibid., 3.

28. Corey to Lechlitner, May 5, 1930.

29. Corey, *Buy an Acre*, 182.

30. Ibid., 183.

31. Ibid., 184.

32. Ibid., 187.

33. Ibid., 188, 189.

34. Mangione, *An Ethnic at Large*, 262.

35. Corey, *Buy an Acre*, 189–195.

36. Lechlitner, "The Last Frontiers," 20.

37. Susman, *Culture as History*.

38. Pells, *Radical Visions*, 36.

39. Doukas, *Worked Over*.

40. Ibid.

41. Susman, *Culture as History*.

42. Doukas, *Worked Over*.

43. Corey, "Lurching toward Liberalism," 52.

44. Susman, *Culture as History*, xxix.

45. Ibid.

46. Hicks, *Where We Came Out*, 106.

47. Ibid.

48. Susman, *Culture as History*, 46.

49. Ibid., xxiv.

50. Lechlitner, "Interview," 20.

51. Isador Schneider to Ruth, October 3, 1935, Ruth Lechlitner Papers.

52. He likewise scorned sloth, as in his 1934 story, "Their Forefathers Were Presidents."

53. Corey, *Five Acre Hill*, 171–172.

54. Corey to Lechlitner, February 17, 1975.

55. Langer, *Josephine Herbst*, 147–149; Pells, *Radical Visions*, 200.

56. Meyer, *The Middle Western Farm Novel*, 94; McCown, "Paul Corey's Mantz Trilogy"; Corey to Arthur W. Wang, February 5, 1964; Corey to Gwyn Kluever, February 10, 1968.

57. Corey, "We Are Americans" (typescript), 41.

58. Mills, *The Sociological Imagination*, 1.

59. Corey to Conroy, September 11, 1933.

60. Corey to Conroy, October 4, 1933.

61. Corey, undated typescript in response to letters from Helen Louise Johnstone, Magazine Liaison Section, Office of International Information & Cultural Affairs, US Office of War Information, February 26, 1946, and March 7, 1946. This typescript is titled "The Mid Westerner" and is filed with correspondence about the *Esprit* story published as "L'homme du Middle West," Paul Corey Papers. Corey used these differences between Danish, Irish, and Yankee farmers as the subject of a 1935 story, "When Farmers Sang."

62. Corey, "Lurching toward Liberalism," 44.

63. Susman, *Culture as History*, xxiii, xxiv.

64. Alexander, *Here the Country Lies*, 29.

65. Susman, *Culture and Commitment*, 5, 6.

66. Ibid., 7.

67. Wixson, *Worker-Writer in America*, 214–219.

68. La Farge, "Foreword," vii–viii.

69. Ibid., vii.

70. McCown, "Paul Corey's Mantz Trilogy."

71. Lechlitner to Spence, 5.

72. Ibid.

73. Kreymborg, Mumford, and Rosenfeld, *The New Caravan*, 163.

74. Ruth to Jerre Mangione and his wife, Patricia, April 4, 1973, Ruth Lechlitner Papers.

Chapter 5: Regionalism and Radicalism

1. Murray, *Red Scare*.

2. Corey, "Lurching toward Liberalism," 53.

3. Rideout, *The Radical Novel*, 132; Pells, *Radical Visions*, 42.

4. Murray, *Red Scare*.

5. Temkin, "Sacco & Vanzetti Today."

6. Cowley, *Exile's Return*, 221.

7. Herbst, "A Year of Disgrace," 96–97.

8. Rideout, *The Radical Novel*, 115.

9. Eisinger, *Fiction of the Forties*, 5–9.

10. Wald, *American Night*, 26, 45.

11. Rideout, *The Radical Novel*, 45–46.

12. Saxton, "Introduction," xviii.

13. Kazin, *On Native Grounds*, 373–380.

14. Strong and Keyssar, *Right in Her Soul*; Mickenberg, *American Girls in Red Russia*.

15. Eisinger, "Character and Self in Fiction," 158.

16. Pells, *Radical Visions*, 14.

17. Durrenberger, "Notes on the Cultural-Historical Background."

18. Wixson, *Worker-Writer in America*, 149.

19. Ibid., 161.

20. Ibid., 174–175 (emphasis original).

21. Johnson, "Introduction," xxiii.

22. Corey, "Lurching toward Liberalism."

23. Ibid., 60.

24. Wixson, *Worker-Writer in America*, 377.

25. Nekola and Rabinowitz, *Writing Red*, 9.

26. Wixson, *Worker-Writer in America*, 377, 378.

27. Parker Wheatley to Corey, January 6, 1940; Corey to Wheatley, January 11, 1940; John F. Frederick to Corey, February 28, 1940.

28. Reigelman, *The Midland*, 46, 49.

29. Ibid., 54 (emphasis original).

30. Ibid.

31. Rideout, *The Radical Novel*; for the distinction between progressives and radicals see Pells, *Radical Visions*.

32. Rideout, *The Radical Novel*, 12–23.

33. Quoted in ibid., 229.

34. Ibid., 209; Madden, "Introduction," xix–xxx.

35. Saxton, "Introduction," xv.

36. Cowley, *Exile's Return*, 4.

37. Wixson, *Worker-Writer in America*, 320, 321. Corey's correspondence with Conroy, Frederick, and Stoll shows clearly that he considered his role as a writer part of a joint enterprise. He was energetic in providing editors with names of people who might subscribe and generous in his contributions.

38. Corey to Conroy, October 4, 1933.

39. Conroy and Johnson, *Writers in Revolt*, x.

40. Ibid., 357.

41. Reigelman, *The Midland*, 16, 17. While the editors of *The Midland* argued they had no support, the University of Iowa paid their wages and provided support in office space and wages for an editorial assistant, the role that Ruth Lechlitner came to Iowa City to fulfill. Furthermore, Reigelman shows that Frederick used his editorship as a negotiating point in setting his salary. Reigelman concludes that while the university's support was indirect, it was nonetheless critical.

42. Corey to Stoll, May 10, 1933.

43. Wixson, *Worker-Writer in America*, 320–321, 290.

44. Suckow, "The Folk Idea in American Life"; Page, "Preface," xxi.

45. Though there may be no direct connection, writings of *The Midland*'s editors, Frederick and John Crowe Ransom, strongly echo this notion of the connection between area, nature, and the people. See Reigelman, *The Midland*, 42; Ransom, "The Aesthetics of Regionalism," 295.

46. Anderson, *Imagined Communities*.

47. Hughes-Hallett, *Gabriele d'Annunzio*; Simeone, "Fascists and Folklorists."

48. Rourke, *The Roots of American Culture*; Hyman, "Constance Rourke."

49. Janson, "The International Aspects of Regionalism."

50. Whiting, *Antifascism*, 99.

51. Lutz, *Cosmopolitan Vistas*, 41, 150; Ellsworth, "Ceres and the American Men of Letters," 179; Hicks, *The Great Tradition*, 282.

52. Meyer, "Introduction," ix.
53. Boulard, *Huey Long Invades New Orleans*.
54. Reigelman, *The Midland*, 65.
55. See, e.g., Durrenberger, *The Dynamics of Medieval Iceland*.
56. Saxton, "Introduction," xv.
57. Ibid., 91–101.
58. For examples, see Suckow, *A Memoir*.
59. Greenleaf, *Fatal Obsession*, 85.
60. Reigelman, *The Midland*.
61. For a summary of Herder, see Björnsdóttir, "The Mountain Woman"; King, *Gods of the Upper Air*, 19.
62. Reigelman, *The Midland*, 40.
63. Ibid., 43.
64. Ibid., 45.
65. Ibid., 3.
66. Ibid., 12, 13.
67. Corey to Wheatley, January 11, 1940; Corey to Schramm, January 9, 1940.

Chapter 6: Naturalism as Ethnography

1. Meyer, "Naturalism in American Farm Fiction"; Meyer cites Parrington, *The Beginnings of Critical Realism*, 232–325. Pells, *Radical Visions*, 338–339.
2. Mayblin, "The Anarchic Institution."
3. Cassuto, *Hard-Boiled Sentimentality*, 4.
4. Brooks, *Painted Horses*, 593.
5. Rahv. "Notes on the Decline of Naturalism," 43.
6. For example, see White, *Islamist Mobilization*; White, *Money Makes Us Relatives*; White, *Muslim Nationalism*; Tuan, *The Good Life*. See also Berger, *Real and Imagined Worlds*, 144–145.
7. Redfield, *Tepoztlan*.
8. Redfield and Warner, "Cultural Anthropology."
9. Van Doren, *Contemporary American Novelists*, 149; Ibid., 153–157.
10. Lewis, "Husbands and Wives," 610. The reference is to Chase's 1931 *Mexico*. In it he compared the idyllic peasant life he observed as a tourist in Mexico, ratified by frequent reference to Redfield's text, with the benighted lives of the clock-driven industrial workers of the Lynds' *Middletown*. As my choice of adjectives shows, he much preferred the rural to the urban/mechanized lives.
11. Burns, *Kinship with the Land*.
12. Manning, "The Trouble with Iowa."

13. Corey, "Corn Gold Farm" (draft; typescript), 247.

14. Goldschmidt, *As You Sow*, 226.

15. Ibid., 227.

16. Ibid., 224, 235, 240.

17. Ibid., 239.

18. Raup, "Corporate Farming."

19. Geertz, *The Interpretation of Cultures*.

20. Clifford and Marcus, *Writing Culture*; Handleman, "Critiques of Anthropology."

21. In the laconic phrasing of D'Arcy McNickle, "Anthropological material [is] normally not subject matter to fascinate a nonprofessional reader." McNickle, *Indian Man*, 222. Sociologist Morroe Berger wrote that "social scientists have introduced fictional techniques to make their work more immediate and palatable." Berger, *Real and Imagined Worlds*, 1.

22. Kroeber, "Introduction."

23. Doukas, *Worked Over*.

24. See, e.g., Boas, "The Methods of Ethnology."

25. Leach, *Rethinking Anthropology*, 2.

26. Kroeber, "Introduction."

27. Clifford and Marcus, *Writing Culture*.

28. Ibid.

29. La Farge, *Raw Material*, 201.

30. Quoted in McNickle, *Indian Man*, 208.

31. Kazin, *On Native Grounds*, 392. See also Pells, *Radical Visions*, 34.

32. Philip Rahv, "Image and Idea," reprinted in Bloom, *American Naturalism*, 39–48, 39.

33. Greenleaf, *Book Case*, 81.

34. Gardiner, *The Art of Fiction*.

35. La Farge, *Raw Material*, 133.

36. Greenleaf, *Book Case*, 13.

37. Kazin, *On Native Grounds*, 368 (emphasis original).

38. Wald, *American Night*, 94.

39. Ibid., 108; Page, "Preface."

40. Elkins, *Old Bones*, 29.

41. For a list of some of these, see Narayan, "Ethnography and Fiction"; Dennis and Aycock, *Literature and Anthropology*.

42. Bandelier, *The Delight Makers*, vi.

43. Kroeber, "Introduction."

44. La Farge wrote other fiction as well, some set in the American Southwest and some in New Orleans. For the converse, an attempt to write without such ethnographic experience, see Fast, *Being Red*, 73.

45. La Farge, *Laughing Boy*, 12.

46. La Farge, *Behind the Mountains*, v.

47. See the review by Gary H. Gossen, "*A Green Tree and a Dry Tree* by Carter Wilson," *American Anthropologist* 76 (1974): 921–923.

48. La Farge, *Raw Material*; McNickle, *Indian Man*, 209; La Farge, *The Man with the Calabash Pipe*, xiii, xiv; Wilson, "Introduction," in *Crazy February*.

49. For more on fiction and ethnography, see Langness and Frank, "Fiction and the Ethnographic Novel"; Frank, "Mercy's Children."

50. Beidler, *Fig Tree John*.

51. Gardner, "Though It Broke My Heart"; Gardner, "Introduction."

52. Glusker, *Anita Brenner*.

53. An engaging treatment of Boas and his students is Charles King's *Gods of the Upper Air*.

54. Hemenway, *Zora Neale Hurston*, 22, 63, 73.

55. Kessler-Harris and Lauter, "Introduction," vii.

56. Hans, *D'Arcy McNickle*.

57. White's ethnography includes *Muslim Nationalism and the New Turks*, *Islamist Mobilization in Turkey*, and *Money Makes Us Relatives*. White's fiction includes *The Sultan's Seal* (2006), *The Abyssinian Proof* (2008), and *The Winter Thief* (2010). See also her 2006 essay, "Portrait of the Scholar as a Young Novelist."

58. Wood's ethnography is *Ogata-Mura*; his novel is *And If Strangers Come to Supper*. Nanda and Gregg, *Assisted Dying*; Nanda and Gregg, *The Gift of a Bride*.

59. Oliver's novels include *Mists of Dawn* (1952), *Shadows in the Sun* (1954), *The Winds of Time* (1956), *Unearthly Neighbors* (1960), *The Wolf Is My Brother* (1967), *The Shores of Another Sea* (1971), *Giants in the Dust* 1976, *Broken Eagle* (1989), and *The Cannibal Owl* (1994). Collections of his short stories include *Another Kind* (1955), *The Edge of Forever* (1971), *A Star Above and Other Stories* (2003), and *Far from This Earth and Other Stories* (2003).

60. Goldschmidt, "Pietro's House."

61. Mintz, "Translator's Note."

62. Benedict, *The Chrysanthemum and the Sword*; Jonsson, "Thai Fiction and the Anthropologist."

63. Fikes, *Carlos Castaneda*.

64. I was interested to learn (316–317) that Berlinski based some elements of the novel on my own ethnographic work.

65. There are many more examples. See, for instance, Wikipedia's "List of Fictional Anthropologists," https://en.wikipedia.org/wiki/List_of_fictional_anthropologists, accessed 1/13/2000.

66. J. Michael Orenduff's Pot Thief novels feature a central character who studied anthropology at the University of New Mexico but was expelled. He laces his conversations with pithy anthropological conclusions.

67. Scott Graham's National Park Mystery series from Torrey House Press, Salt Lake City. These include *Canyon Sacrifice* (2014), *Mountain Rampage* (2015), *Yellowstone Standoff* (2016), *Yosemite Fall* (2018), *Arches Enemy* (2019), *Mesa Verde Victim* (2020), and *Canyonlands Carnage* (2021).

68. Kazin, *On Native Grounds*; Hicks, *The Great Tradition*, 179–180; Pells, *Radical Visions*, 248.

69. Redfield, *Tepoztlan*.

70. Chase, *Mexico*; Britton, *Revolution and Ideology*, chap. 7; Pells, *Radical Visions*, 101.

71. Beals, "Aboriginal Survivals in Mayo Culture."

72. Lewis, "Husbands and Wives," 610.

73. Susman, *Culture as History*, 156.

74. Lewis, *Life in a Mexican Village*.

75. La Farge, "Foreword," vii–viii.

76. Ibid., ix.

77. Reigelman, *The Midland*, 65.

78. For instance, his award-winning *La Vida* (1966).

79. Quoted in Solotaroff, *Robert Stone*, ix.

80. Eisinger, "Character and Self," 160.

81. Pells, *Radical Visions*, 229. For an anthropological view, see Durrenberger and Erem, "The Weak Suffer What they Must."

82. Frank Nein to Corey, n.d., filed with "Number two head-saw" drafts and correspondence. Howard Fast was another American writer who appreciated the relationship between experience and writing. See Fast, *Being Red*, especially 71–73.

83. Ibid.

84. McCown, "Paul Corey's Mantz Trilogy"; Gerber, "Becoming an Author," 11; Corey to Marshall Best, March 10, 1938.

85. Wixon, *Worker-Writer in America*, 332.

86. Lynd and Lynd, *Middletown*, vi.

87. Goldschmidt, *As You Sow*, vii.

88. Buel Beems to Corey, 1935.

89. Larson, Williams, and Wimberley, "Dismissal of a Sociologist."

90. Wallace, "Foreword," v.

91. Miner, *The Primitive City of Timbuctoo.*

92. Miner, *Culture and Agriculture.*

93. Ibid., iii–iv.

94. Shockley, "The Reception of *The Grapes of Wrath*"; Butterworth, "Oscar Lewis."

95. Friedland, "Who Killed Rural Sociology?"

96. Larson and Zimmerman, "The USDA's Bureau of Agricultural Economics," 238.

97. Kirkendall, *Social Scientists and Farm Politics,* 224.

Chapter 7: The Demise of Iowa's Family Farms

1. In the collection of Corey's papers at the University of Iowa Library, one folder is labeled "Three Miles Square." A typed note, probably by Corey, says, "Correspondence from the beginning to publication and after. The author's various attempts to get editors [*sic*] interest. The record of cutting the manuscript for publication. Reviews and advertisements, and other items of publicity. The Railway Express record of manuscript transportation. Some detailed letters of criticism of the book before and after publication." And perhaps more importantly, the folder contains Corey's own summaries of the book and his conceptual scheme for the series of books he planned to write. And there is more correspondence regarding *Three Miles Square* between Corey and editors and his agent as well as others in the correspondence files by date. There are two versions of the overall plan. I shift between them depending on which has the fuller treatment.

2. Corey, *Three Miles Square,* 1–2.

3. "Summary of Novel Program" (typescript), 1.

4. Ibid.

5. Ibid., 2–3.

6. Ibid., 3.

7. Corey, "Presentation of: Manuscript 'Three Miles Square'; Author Paul Corey; Author's Writing Program" (transcript), 2.

8. Dyson, *Farmers' Organizations,* 84.

9. Ibid.

10. White, *Milo Reno,* 10.

11. Ibid.; Dyson, *Farmers' Organizations,* 85.

12. Dyson, *Farmers' Organizations,* 85.

13. Dyson, *Red Harvest,* 102–103.

14. Federal Writers' Project, *Iowa,* 305.

15. Ibid., 75–77.

16. Karr, "Farmer Rebels," 637.

17. Dyson, *Red Harvest*, 74–76.

18. Ibid., 80.

19. Karr, "Farmer Rebels."

20. Corey, "Lurching toward Liberalism," 66.

21. Federal Writers' Project, *Iowa*, 60.

22. Corey, "We Are Americans" (typescript), 3.

23. Wertsch, "Resisting the Wave," 1–2.

24. Ibid., 2.

25. Ibid., 143.

26. Ibid., 145.

27. Ibid.

28. Ibid., 146.

29. Ibid., 204–205.

30. Ibid., 206.

31. Ibid., 208.

32. Corey, "We Are Americans" (typescript), 3–4.

33. Handwritten synopsis, filed with the typescript of the first chapters of a work titled "We Are Americans."

34. Corey, "We Are Americans" (typescript), 1–2.

35. Ibid., 3.

36. Ibid., 43.

37. Corey, "Presentation of: Manuscript 'Three Miles Square'; Author Paul Corey; Author's Writing Program" (transcript), 4.

38. Biser, *Cooperative Supply and Equipment Operations*.

39. Corey, "We Are Americans" (typescript), 58–59.

40. "WFEA Books of the Week" (typescript), Thursday, September 5, 1946.

41. "Summary of Novel Program" (typescript), 5.

Chapter 8: The One or the Many?

1. Quoted in Gerber, "Becoming an Author," 21.

2. Burns, *Kinship with the Land*.

3. Meyer, *The Middle Western Farm Novel*, 98.

4. Wixon, *Worker-Writer in America*, 214, 219, 220.

5. Quoted in Meyer, *The Middle Western Farm Novel*, 96.

6. Corey, "Lurching toward Liberalism," 41.

7. "Green Lumber" (typescript), 4.

8. This is probably after Jack Reed, the journalist and communist who wrote *Ten Days that Shook the World*, about the Bolshevik revolution.

9. "Green Lumber" (typescript), 29.

10. Ibid., 43.

11. Ibid., 43–50.

12. Corey to Stoll, July 10, 1933. Malcolm Cowley identified writing about groups rather than individuals except insofar as they exemplify groups as characteristic of naturalist writing. Cowley, "Naturalism in American Literature," 75.

13. Corey to O'Brien, November 27, 1934.

14. Corey to Frances Phillips, William Morrow & Co., March 31, 1937.

15. Eisinger, *Fiction of the Forties*, 64.

16. Kazin, *On Native Grounds*, 381–382.

17. Kazin, *Starting Out in the Thirties*, 52–54.

18. Cowley, "Naturalism in American Literature," 79.

19. Wald, *American Night*, 94.

20. Maxim Lieber to Corey, April 10, 1935.

21. Rosenfeld to Corey, May 6, 1936.

22. John H. Thompson to Corey, July 18, 1936.

23. William James Fadiman to Corey, October 14, 1936.

24. Fadiman to Corey, November 4, 1936.

25. Martha Foley to Corey, July 13, 1936, and July 30, 1936.

26. Helen Lincoln to Corey, November 11, December 4, and December 15, 1936.

27. Frances Phillips to Corey, December 10, 1936, January 6, 1937.

28. Gorham Munson to Jerre Mangione, January 20, 1937.

29. Conroy to Corey, February 25, 1933; Corey to Conroy, April 5, 1933.

30. Conroy to Corey, March 12, 1935.

31. Corey to Stoll, April 5, 1933; Corey to Conroy, April 22, 1935.

32. Gerber, "Becoming an Author."

33. Beems to Corey, March 14, 1935; Beems to Corey, 1935.

34. Beems to Corey, 1935.

35. Marshall Best to Corey, March 18, 1938.

36. Mavis McIntosh to Corey, October 14, 1938. McIntosh is quoting from a letter to her from Lambert Davis, the Bobbs Merrill editor.

37. McIntosh to Corey, October 14, 1938.

38. Corey to McIntosh, October 16, 1938.

39. McIntosh to Corey, October 21, 1939.

40. Gerber, "Becoming an Author," 8.

41. Corey to Wang, February 5, 1964.

42. Flanagan, "The Middle Western Farm Novel," 121.

43. Corey to Edward J. O'Brien, November 27, 1934.

44. Conroy to Corey, January 21, 1934.

45. Rahv to Corey, February 18, 1935.

46. Corey to Rahv, March 6, 1935.

47. Undated typescript Corey submitted to several magazines in 1935 and 1936.

48. Whitehead, *John Lehmann's "New Writing."*

49. Shover, *Cornbelt Rebellion*; Vogeler, *The Myth of the Family Farm*; Pratt, "Rethinking the Farm Revolt."

50. Arnold Gingrich to Corey, n.d.

51. Editors of *New Masses* to Corey, n.d.

52. Marion Ives to Corey, September 27, 1935.

53. Paul Rumfeld for W. W. Norton Editors to Corey, January 4, 1936.

54. Rumfeld to Corey, January 23, 1936.

55. W. W. Norton to Corey, October 20, 1936; Rumfeld to Corey, October 9, 1936.

56. Ibid.

57. Fadiman to Corey, November 9, 1936.

58. Helen Lincoln to Corey, December 9, 1936.

59. LeSeur to Corey, n.d.

60. Corey to LeSeur and [Dale] Kramer, n.d.

61. Ibid.

62. Corey to Gipson, April 19, 1937.

63. Corey to Field, May 5, 1937.

64. Field to Corey, May 20, 1937.

65. Alan Calmer (International Publishers letterhead) to Corey, July 23, 1937.

66. Corey to Calmer, July 23, 1937.

67. Field to Corey, July 29, 1937.

68. Calmer to Corey, July 1937.

69. Ibid.

70. Ibid.

71. Corey to Calmer, August 1, 1937.

72. Corey to Herb, September 9, 1937.

73. Field to Corey, September 12, 1937.

74. Corey to Field, September 28, 1937.

75. Field to Corey, September 29, 1937.

76. Field to Corey, October 3, 1937.

77. Hirsch to Corey, October 8, 1937; Corey to Hirsch, September 16, 1937.

78. Auerbach to Corey, October 18, 1937.

79. Field to Corey, November 14, 1937.

80. Corey to Auerbach, October 22, 1937.

81. Corey to Henderson, January 4, 1938; Howe to Corey, February 21, 1938; Conklin to Corey, March 12, 1938; Best to Corey, March 21, 1938; Burnett to Corey, March 29, 1939; Conroy to Corey, May 5, 1939.

82. Best to Corey, April 5, 1938.

83. Davis to Corey, November 4, 1940.

84. Ibid.

85. McIntosh to Corey, November 7, 1940.

86. Davis to Corey, May 7, 1941.

87. Davis to Corey, January 27, 1942.

88. Corey to McIntosh, May 17, 1942.

89. Hicks to Corey, June 26, 1942.

90. Hicks to Corey, July 9, 1942.

Chapter 9: Agents and Editors

1. E. G. Rice to Corey, January 30, 1933.

2. Rice to Corey, October 4, 1932.

3. Corey to Rice, October 6, 1932.

4. Corey and Lechlitner met D. H. Lawrence at Bandol in France. Corey, "Lurching toward Liberalism," 54.

5. Corey to Rice February 2, 1933.

6. Corey to Rice February 2, 1933.

7. Rice to Corey, February 23, 1933.

8. Maxim Lieber to Corey, June 16, 1933; Lieber to Corey, April 10, 1935.

9. Mavis McIntosh to Corey, August 3, 1938.

10. Conroy to Corey, April 19, 1933.

11. *Company K* was a 1933 novel by William March. It was serialized in the magazine *The Forum* before Smith and Haas published the complete novel. It consists of vignettes from the points of view of 113 Marines in the company.

12. Corey to Conroy, April 22, 1933.

13. Corey to Conroy, January 13, 1934.

14. Corey to Stoll, Apr 5, 1933.

15. Wertsch, "Resisting the Wave."

16. Caldwell, *God's Little Acre*, 56–57.

17. Corey, *Three Miles Square*, 149.

18. Corey, *The Road Returns*, 112.

19. See, e.g., King, *Southern Ladies and Gentlemen*.

20. Nelson, "My Uncle Paul," 57.

21. Stoll to Corey, August 2, 1933. About a month later Corey received a letter from Jack Conroy confirming the story. Conroy to Corey, September 7, 1933.

22. Stoll to Corey, August 29, 1933.

23. Uzzell to Corey, September 14, 1933.

24. Stoll to Corey, September 18, 1933.

25. Corey to Uzell, September 20, 1933.

26. Corey to Uzell, September 20, 1933.

27. Corey to Uzzell, September 9, 1933. Corey's reference to a "mental breakdown" is the first and only mention of such an event that I have found.

28. Corey to Uzzell, December 9, 1933.

29. Corey to Uzzell, November 4, 1933.

30. Uzzell to Corey, December 30, 1933.

31. Paul Corey, "Summary of the novel: Acre March" (typescript), [1933?].

32. Ibid.

33. Uzzell to Corey, January 26, 1934.

34. Corey to Uzzell, March 3, 1934; Corey to Uzzell, June 7, 1934.

35. Gerber, "Becoming an Author," 9.

36. Corey to Uzzell, March 3, 1934; Lechlitner, ". . . anti-war and anti-fascism . . . ," 80.

37. Nowell to Corey, April 28, 1937.

38. Corey to Nowell, May 1, 1937.

39. Nowell to Corey, May 9, 1937.

40. Corey to Nowell, May 10, 1937.

41. Kerker Quinn to Corey, May 11, 1935.

42. Corey to Quinn, February 20, 1935.

43. Corey to Richard Perry, March 9, 1935.

44. Ibid.; Corey to Marshall A. Best, March 10, 1938; Corey to Nowell, 1938, 11. The reference is to the anthropological study of Muncie, Indiana, by Robert and Helen Lynd. The town's chief industry was the production of Mason jars, widely used for canning.

45. Corey to Henningsen, October 15, 1945.

46. Gerber, "Becoming an Author," 12.

47. Ibid., 17, 18.

48. Mangione, *The Dream and the Deal*, 4.

49. Ibid., 9.

50. Ibid.

51. Ibid., 217.

52. Ibid., 290.

53. Ibid., 150.

54. Ibid., 149, 150.

55. Ibid., 150.

56. Ibid.

57. Ibid.; Corey to Monty Noam Penkower, February 19, 1968. I queried the FBI under the Freedom of Information Act, but the bureau responded that it had no file on Corey or Lechlitner.

58. Lechlitner to Corey, Thursday AM.

59. Lechlitner to Corey, Saturday AM.

60. Quoted in Mangione, *The Dream and the Deal*, 150.

Chapter 10: A Collective Story

1. Rahv to Corey, February 9, 1935.

2. Corey to Rahv, February 9, 1935.

3. Eisinger, "Character and Self in Fiction," 182.

4. W.W.B. for the Editors of *New Masses* to "Contributor," n.d. The note was probably sent in 1935, about the same time Corey was looking for a venue to republish "Bushel of Wheat, Bushel of Barley." See chap. 8.

5. Joseph Gollomb to Corey, August 10, filed with "No. Two Head Saw."

6. M. Tjader Harris to Corey, July 16, filed with "No. Two Head Saw."

7. John J. Trounstine to Corey, December 30, 1936.

8. *Critics Group* to Corey, August 18, 1938.

9. Corey to Ben (of the *Critics Group*), August 20, 1938.

10. *Critics Group* to Corey, September 6 [1938].

11. Wilkening to Corey, February 19, 1940.

12. Corey, "Number Two Head-Saw" (transcript). Corey started submitting it in 1935 and continued for the next five years.

Chapter 11: Paul's War

1. Mangione, *An Ethnic at Large*, 262.

2. R. W. Kingman to Corey, April 28, 1937.

3. Corey to Commanding Officer, Second Corps Area, Governors Island, New York, July 5, 1940.

4. Chalmers Dale to Corey, August 15, 1941.

5. Paul Corey, Putnam County Defense Corps ID card.

6. Corey to Dale, December 19, 1941.

7. Dale to Corey, December 22, 1941.

8. Corey to Dale, December 26, 1941.

9. MacLeish to Corey, January 8, 1942.

10. GET from RL 1973. . . . w/macleish. Lechlitner to Carroll Coleman, March 10, 1973.

11. War Department to Corey, February 7, 1942.

12. Corey to Wallace, April 29, 1942; Wallace to Corey, May 6, 1942.

13. Corey to Mead, March 20, 1943.

14. Corey to James C. Patton, March 28, 1943.

15. Paul W. White to Corey, June 10, 1943.

16. Pierce A. Cumings to Corey, July 13, 1943.

17. Benson, *John Steinbeck*, 511–512, 513.

18. Winks, *Cloak and Gown*.

19. PM to Corey, December 16, 1941.

20. Corey to Luce, December 22, 1941.

21. Ann R. Elgar to Corey, December 31, 1941.

22. Corey to Mavis McIntosh, June 23, 1942.

23. Charles Burns to Corey, May 25, 1942.

24. Corey to James N. Mead, June 17, 1942.

25. Corey to McIntosh, June 23, 1942.

26. Corey to Daniel D. Mich, June 23, 1942; Mich to Corey, June 25, 1942.

27. Burns to Corey, July 7, 1942.

28. Corey, "Hitler Can Invade America by Air."

29. Bird to Corey, September 23, 1942.

30. Corey to Bird, September 27, 1942.

31. Corey to Homer and Goldie, August 20, 1987. The reference is to Harvey Hollister Bundy, who served as a special assistant to Secretary of War Henry L. Stimpson. In a letter dated March 30, 1989, to Mel Yoken, professor of French at the University of Massachusetts, Dartmouth, Corey mentions the story and explains that the reference was to a special assistant to Stimpson, a person named Bundy, whose first name he thought was Avery and who was a friend of Bird's. He also wrote that Bird had hired him to write the story. Corey, Bird, and Lechlitner referred to this story as "Mr. Bundy's Horror Story" in their correspondence. Corey later settled on the title "For Want of a Nail." There is no evidence in the Corey Papers to suggest that it was ever published.

32. Corey, "For Want of a Nail" (typescript), n.d. (probably 1941 or 1942).

33. Bird to Corey, September 29, 1942, October 2, 1942; H. Wendell Endicott to Corey, October 9, 1942.

34. Bird to Corey, October 13, 1942.

35. Corey to Bird, October 20, 1942.

36. Annie Laurie Williams to Corey August 28, 1942.

37. Corey to Mavis McIntosh, June 23, 1942.

38. Helen Strauss to Jerre Mangione, September 10, 1942.

39. Henry A. Wallace, "The Century of the Common Man," box 50, Henry A. Wallace Papers, Special Collections, UIL. A frequently reprinted speech delivered to the Free World Association at the Grand Ballroom of the Commodore Hotel, New York, May 8, 1942.

40. Corey, "The American People Can Fight" (typescript), n.d.

41. Rankin, *Telegram from Guernica*, 115.

42. Wintringham, "Introduction," reprinted in Marxists Internet Archive Library, https://www.marxists.org/archive/wintringham/1941/x01/intro.htm, accessed 6/6/21.

43. Osanka, "Introduction."

44. Corey to Lechlitner, October 19, 1942.

45. Corey to Lechlitner, October 21, 1942.

46. Corey to Lechlitner, October 24, 1942.

47. Bird to Corey, October 21, 1942; Bird to Corey (on Somerset Club stationary), n.d.

48. Corey to Bird, October 30, 1942.

49. Bird to Corey, November 2, 1942.

50. Bird to Corey, November 3, 1942.

51. Corey to Bird, November 6, 1942.

52. Corey to Bird, November 7, 1942.

53. Bird to Corey, November 9, 1942.

54. Corey to Bird, November 12, 1942.

55. Bird to Corey, November 17, 1942.

56. Corey to Bird, November 18, 1942.

57. Corey to Bird, November 19, 1942.

58. Corey to Bird, November 30, 1942.

59. Bird to Corey, December 7, 1942.

60. Corey to Bird, December 9, 1942.

61. Bird to Corey, December 10, 1942.

62. Wolfskill and Hudson, *All but the People*, 85, 160–166, 304–308; Phillips-Fein, *Invisible Hands*, 10–13, 20–22.

63. Corey to Bird, December 14, 1942.

64. Bird to Corey, December 17, 1942; strikeout is in the original.

65. Corey to Bird, December 21, 1942.

66. Bird to Corey, December 24, 1942; Corey to Bird, December 30, 1942; Bird to Corey, January 5, 1943.

67. Corey to Bird, January 11, 1943.

68. Bird to Corey, January 13, 1943.

69. Corey to Bird, March 2, 1943, March 9, 1943.

70. Ibid., 22.

71. Ibid., 33.

72. Ibid., 46.

73. Ibid., 50.

74. Ibid., 51.

75. Karl (at *Look*) to Corey, September 9, 1942.

76. Robert Zellers to Corey, August 25, 1942.

77. Nugent M. George to Corey, August 27, 1942.

78. Cited in Kenneth V. Summer to the editor, *Look*, September 10, 1942.

79. Harry. A. Ware to editorial office, *Look* (handwritten letter), September 5, 1942.

80. H. E. Clow to editor, *Look*, September 14, 1942.

81. Leonard W. Cronkhite to Corey, October 16, 1942.

82. Judge Malcolm Hatfield to Corey, November 7, 1942.

83. T.A.H. Taylor to editor, *Look*, September 17, 1942; "Karl" (at *Look*) to Corey, September 9, 1942.

84. Wolfskill and Hudson, *All but the People*, 61.

85. Ibid.

86. Schlesinger, *The Politics of Upheaval*, 79–95; Archer, *The Plot to Seize the White House*, 21.

87. Strong, *My Native Land*, chap. 13.

88. Pells, *Radical Visions*, 78.

89. Wolfskill, *The Revolt of the Conservatives*.

90. La Farge, *Raw Material*, 103.

91. Denton, *The Plots against the President*, 145; Culver and Hyde, *American Dreamer*, 115–116.

92. Culver and Hyde, *American Dreamer*, 51. See also Saxton, *Grand Crossing*, 143.

93. Wolfskill, *The Revolt of the Conservatives*, 19.

94. Archer, *The Plot to Seize the White House*, 21.

95. Corey to Luce, December 22, 1941.

96. Archer, *The Plot to Seize the White House*, 21.

97. Schlesinger, *The Politics of Upheaval*, 518–523; Archer, *The Plot to Seize the White House*, 31.

98. Denton, *The Plots against the President*, 25.

99. Strong, *My Native Land*, 185–189.

100. Boulard, *Huey Long*.

101. Denton, *The Plots against the President*, 215.

102. Fast, *Being Red*, 95.

103. Denton, *The Plots against the President*, 171–175.

104. Schlesinger, *The Politics of Upheaval*, 82, 83.; Denton, *The Plots against the President*, 175; Archer, *The Plot to Seize the White House*; Wolfskill, *The Revolt of the Conservatives*.

105. Schlesinger, *The Politics of Upheaval*; Archer, *The Plot to Seize the White House*.

Chapter 12: Ruth's War

1. Thompson and Winnick, *Robert Frost*.

2. Lechlitner, "Biographical," 3.

3. Mangione, *An Ethnic at Large*, 262.

4. Crystal Eastman, quoted in Roberts, *Three Radical Women Writers*, 4.

5. La Farge, *Raw Material*, 209–210.

6. Lechlitner to Dr. Anne Spence, titled "For Dr. Anne Spence," n.d. (probably 1975), 8, 9. Lechlitner wrote this during what was obviously a contentious period between her and Paul. Some of Corey's statements are not borne out by other evidence, and it would not be appropriate to treat this or his writings regarding the same matter as authoritatively accurate.

7. Ibid., 7, 8.

8. Maas to Lechlitner, November 24, 1935.

9. Graham, "Mystery Man."

10. Morse, "Review of *Tomorrow's Phoenix*," 157.

11. For a discussion of this see Pells, *Radical Visions*, 152–153.

12. Lechlitner, "Ordeal by Tension," 296–299.

13. La Farge, *Raw Material*, 1.

14. Lechlitner, "Quiz Program," 132–133.

15. Ibid., 132–133.

16. Kaplan, "American Speech in Radio Poetry"; Brunner, *Cold War Poetry*, 189.

17. Stein, "Industrial Music," 244–245.

18. Lechlitner to Decker, January 31, 1945.

19. Kreymborg to Lechlitner, April 12, 1945.

20. Mary Squire Abbot (of McIntosh & Otis, agents) to Mrs. Corey, November 9, 1945.

21. See Harrison, "Confirmed Typomaniac."

22. Coleman to Lechlitner, March 4, 1973.

23. Hallwas, "Poetry and Murder."

24. Brunner, *Cold War Poetry*, 188–189.

25. Lechlitner, "Lines for the Year's End," 72–73.

Chapter 13: California

1. Mangione to Paul and Ruth, January 23, 1946.

2. Huff, "Living High on $6500 a Year," 60.

3. Corey, "We Bought an Acre" (typescript), n.d. (probably 1963 or earlier), 5; Lechlitner to Conroy, May 9, 1973.

4. Corey, *Homemade Homes*, 176–177.

5. Ibid., 177.

6. Corey to Lechlitner, May 1, 1946.

7. Ibid.

8. Corey to Lechlitner, May 3, 1946.

9. Corey to Lechlitner, May 6, 1946; May 7, 1946; May 8, 1946.

10. Corey to Lechlitner, May 10, 1946; May 11, 1946; May 12, 1946; May 15, 1946; May 17, 1946; May 18, 1946.

11. Corey to Lechlitner, May 20, 1946.

12. Corey to Lechlitner, May 21, 1946.

13. Paul Corey to Anne Corey, May 22, 1946.

14. Corey to Lechlitner, May 22, 1946.

15. Corey to Lechlitner, May 30, 1946.

16. DeJong to Lechlitner and Corey, June 21, 1948.

17. Ibid.

18. Corey to R. E. Ellsworth, January 2, 1946.

19. Corey, *Homemade Homes*, 181.

20. Fran Huff to Coreys, July 18, 1946.

21. Fran Huff to Coreys, July 20, 1946.

22. Darryll Huff to Ruth, Paul, and Anne, July 1946.

23. Fran Huff to Coreys, Monday night 8 p.m.

24. Fran Huff to Coreys, August 1, 1946; August 2, 1946.

25. Darryll Huff to Corey, August 9, 1946.

26. Fran Huff to Coreys, August 18, 1946.

27. Fran Huff to Coreyios, August 25, 1946.

28. Fran Huff to Coreys, September 9, 1946; September20, 1946.

29. Zwart to Corey, September 21, 1946.

30. Fran Huff to Coreys, October 2, 1946.

31. Darryll Huff to Coreys, October 9, 1946; Darrell and Fran Huff to Coreys, October 11, 1946.

32. Fran Huff to Coreys, November 9, 1946.

33. Huff, "Living High on $6500 a Year," 61.

34. Lechlitner to Carroll Coleman, March 10, 1973; Corey, *Homemade Homes*, 182.

35. Corey, "Fugitives from Gray Flannell [*sic*] or We Bought an Acre" (typescript), 5, 7.

36. Corey, *Homebuilt Homes*, 188, 229, 231, 214.

37. Lechlitner to Cowley, April 30, 1964. The book *A Changing Season: Selected and New Poems, 1962–1972* was published in 1973 after nearly a decade of correspondence with various publishers and constant rejection from them.

38. Ibid.

39. Winnick, "Preface," xi, xii.

40. Lechlitner to Corwin, April 22, 1964.

41. Ibid.

42. Corwin to Lechlitner, April 28, 1964.

43. Corey to Lechlitner, May 12, 1964. It was the practice to enclose a stamped, self-addressed envelope with any submission for the return of the manuscript and letter of rejection.

44. Rexroth to Ruth Corey, May 12, 1964.

45. Lechlitner to Corey, May 20, 1964.

46. Corey to Lechlitner, May 22, 1964

47. Lechlitner to Corey, May 26, 1964.

48. Anne Corey to Paul Corey, May 25, 1964.

49. Ibid.

50. Thompson to Lechlitner, July 17, 1964.

51. Lechlitner, *A Changing Season*, 40–41.

52. Lechlitner to Thompson, October 12, 1964.

53. Ibid.

54. Lechlitner to Thompson, February 12, 1967. "Birchite" refers to a member of the John Birch Society, the right-wing anti-communist organization.

55. Corey to Gwyn Kluever, February 10, 1968.

56. Lechlitner to Thompson, June 6, 1971.

57. Lechlitner, *A Changing Season*, 53–55. "America's Greening" is a reference to Charles A. Reich's *The Greening of America* (1970), which predicted a nonviolent revolution that would occur with the transformation of individuals by a change in consciousness that he called Consciousness III.

58. Thompson to Lechlitner, September 16, 1970.

59. Olsen, *Tell Me a Riddle*, 33.

60. Corey to Lechlitner, February 17, 1975.

61. Lechlitner to Dr. Anne Spence, n.d. [ca. 1975], 1; much of the same information is in Lechlitner's marginal notes in *The Sexual Responsibility of Women* (1956) by Maxine Davis; Corey, "The Pattern" (typescript), n.d. [ca. 1975].

62. Wixson, "Looking for Paul Corey," 68.

63. Corey to Lyn, January 9, 1988.

64. Corey to Homer and Goldie, March 7, 1988.

65. Corey to Lyn, July 6, 1988.

66. Corey to Marie Nelson, August 20, 1988.

67. Corey to Margaret Nelson, October 18, 1988.

68. Corey to Lyn, November 28, 1988.

69. Corey to Mel Yoken, August 14, 1989.

70. Wixson, "Looking for Paul Corey," 70.

Epilogue

1. Federal Writers' Project, *Iowa*, 29.

2. "I wasn't able to write about Iowa until I came here to New York," Corey wrote to his high school English teacher. Corey to Henningsen, October 15, 1945.

3. Wixson, "Looking for Paul Corey," 67.

4. Manning, "The Trouble with Iowa."

BIBLIOGRAPHY

Alexander, Charles C. *Here the Country Lies: Nationalism and the Arts in Twentieth Century America*. Bloomington: Indiana University Press, 1980.

Anderson, Benedict. *Imagined Communities: Reflections on the Origin and Spread of Nationalism*. London: Verso, 1983.

Anderson, Stephanie. *One Size Fits None: A Farm Girl's Search for the Promise of Regenerative Agriculture*. Lincoln: University of Nebraska Press, 2019.

Archer, Jules. *The Plot to Seize the White House*. New York: Hawthorne Books, 1973.

Bandelier, Adolph. *The Delight Makers*. New York: Dodd, Mead, 1890.

Banner, Lois W. *Intertwined Lives: Margaret Mead, Ruth Benedict, and Their Circle*. New York: Alfred A. Knopf, 2003.

Beals, Ralph L. "Aboriginal Survivals in Mayo Culture." *American Anthropologist* 34, no. 1 (1932): 28–39.

Behar, Ruth. *Translated Woman: Crossing the Border with Esperanza's Story*. Boston: Beacon Press, 1993.

Beidler, Peter G. *Fig Tree John: An Indian in Fact and Fiction*. Tucson: University of Arizona Press, 1977.

DOI: 10.5876/9781646422081.c015

Belfrage, Cedric. *The American Inquisition: 1945–1960.* Indianapolis: Bobbs-Merrill, 1973.

Benedict, Ruth. *The Chrysanthemum and the Sword: Patterns of Japanese Culture.* New York: Houghton Mifflin, 1946.

Benedict, Ruth. *Patterns of Culture.* Boston: Houghton Mifflin, 1934.

Benson, Jackson J. *John Steinbeck, Writer: A Biography.* New York: Penguin, (1984) 1990.

Benson, Jackson J. *Wallace Stegner: His Life and Work.* New York: Viking, 1996.

Berger, Morroe. *Real and Imagined Worlds: The Novel and Social Science.* Cambridge, MA: Harvard University Press, 1977.

Berger, Samuel R. *Dollar Harvest: The Story of the Farm Bureau.* Lexington, MA: Heath Lexington Books, 1971.

Berlinski, Mischa. *Fieldwork: A Novel.* New York: Farrar, Straus and Giroux, 2007.

Bernard, H. Russell. *Research Methods in Anthropology: Qualitative and Quantitative Approaches.* New York: AltaMira, 2011.

Bindas, Kenneth. "Review of Michael Denning, *The Cultural Front: The Laboring of American Culture in the Twentieth Century.*" *American Historical Review* 103, no. 3 (1998): 996–997.

Biser, Lloyd C. *Cooperative Supply and Equipment Operations: Farmer Cooperatives in the United States.* Cooperative Information Report 1. Section 20. Washington, DC: US Department of Agriculture, Agricultural Cooperative Service, 1980.

Björnsdóttir, Inga Dóra. "The Mountain Woman and the Presidency." In *Images of Contemporary Iceland*, edited by Gísli Pálsson and E. Paul Durrenberger, 106–125. Iowa City: University of Iowa Press, 1996.

Blackman, Margaret B. "The Individual and Beyond: Reflections on the Life History Process." *Anthropology and Humanism Quarterly* 16, no. 2 (1991): 56–62.

Block, W. J. *The Separation of the Farm Bureau and the Extension Service.* Urbana: University of Illinois Press, 1960.

Bloom, Harold. 2004. *American Naturalism.* Broomall, PA: Chelsea House, 2004.

Boas, Franz. "The Methods of Ethnology." *American Anthropologist* 22, no. 4 (1920): 311–321.

Bohannan, Laura. *Return to Laughter.* New York: Harper, 1954.

Borsodi, Ralph. *Flight from the City: An Experiment in Creative Living on the Land.* New York: Harper and Row, 1933.

Boulard, Garry. *Huey Long Invades New Orleans: The Siege of a City, 1934–1936.* Gretna, LA: Pelican, 1998.

Bowers, William L. *The Country Life Movement in America: 1900–1920.* Port Washington, NY: Kennikat Press, 1974.

Brooks, Malcolm. *Painted Horses.* New York: Grove Press, 2014.

Britton, John A. *Revolution and Ideology: Images of the Mexican Revolution in the United States.* Lexington: University Press of Kentucky, 1995.

Brunner, Edward. *Cold War Poetry.* Urbana: University of Illinois Press, 2001.

Burns, E. Bradford. *Kinship with the Land: Regionalist Thought in Iowa, 1894–1942.* Iowa City: University of Iowa Press, 1996.

Busch, L., ed. *Science and Agricultural Development.* Totowa, NJ: Allanheld, Osmun, 1981.

Butterworth, Douglas. "Oscar Lewis 1914–1970." *American Anthropologist* 74, no. 3 (1972): 747–757.

Caldwell, Erskine. *God's Little Acre.* New York: Viking, 1933.

Canovan, M. *Populism.* New York: Harcourt Brace Jovanovich, 1981.

Cassuto, Leonard. *Hard-Boiled Sentimentality: The Secret History of American Crime Stories.* New York: Columbia University Press, 2009.

Chase, Stuart. *Mexico: A Study of Two Americas.* New York: Macmillan, 1931.

Chayanov, A. V. *The Theory of Peasant Economy.* Edited by D. Thorner, B. Kerblay, and R.E.F. Smith. Homewood, IL: American Economic Association, 1966.

Clifford, James, and George E. Marcus. *Writing Culture: The Poetics and Politics of Ethnography.* Berkeley: University of California Press, 1986.

Cochrane, W. W. *The Development of American Agriculture.* Minneapolis: University of Minnesota Press, 1979.

Coiner, Constance. *Better Red: The Writing and Resistance of Tillie Olsen and Meridel Le Sueur.* New York: Oxford University Press, 1995.

Collins, Robert M. "American Corporatism: The Committee for Economic Development, 1942–1964." *The Historian* 44. no. 2 (February 1982): 151–173.

Conklin, Paul K. *Tomorrow a New World: The New Deal Community Program* Ithaca, NY: Cornell University Press, 1959.

Conroy, Jack, and Curt Johnson, eds. *Writers in Revolt: The Anvil Anthology.* New York: Lawrence Hill, 1973.

Corey, Paul. *Acres of Antaeus: A Novel about Farms and Farm Empires.* New York: Henry Holt, 1946.

Corey, Paul. "Bridge over the Grand Vallat." *Blue Book Magazine* 93, no. 3 (1951): 91–97.

Corey, Paul. "Bushel of Wheat; Bushel of Barley." In *The New Caravan,* edited by Alfred Kreymborg, Lewis Mumford, and Paul Rosenfeld, 364–409. New York: W. W. Norton, 1936.

Corey, Paul. *Buy an Acre: America's Second Front.* New York: Dial Press, 1944.

Corey, Paul. *Corn Gold Farm.* New York: William Morrow, 1948.

Corey, Paul. *County Seat.* Indianapolis: Bobbs-Merrill, 1941.

Corey, Paul. *Five Acre Hill.* New York: William Morrow, 1945.

Corey, Paul. "Hitler Can Invade America by Air." *Look* 6, no. 18 (September 1942): 13–17.

Corey, Paul. *Homemade Homes.* New York: William Sloane Associates, 1950.

Corey, Paul. *The Little Jeep.* Cleveland: World Publishing Company, 1946.

Corey, Paul. "Lurching toward Liberalism: Political and Literary Reminiscences." *Books at Iowa* 47, no. 1 (1988): 35–71.

Corey, Paul. *Red Tractor.* New York: William Morrow, 1944.

Corey, Paul. *The Road Returns.* Indianapolis: Bobbs-Merrill, 1940.

Corey, Paul. *Three Miles Square.* Indianapolis: Bobbs-Merrill, 1939.

Corey, Paul. "Their Forefathers Were Presidents." *Story* 5, no. 24 (July 1934): 56–64.

Corey, Paul. "When Farmers Sang." *Story* 7, no. 37 (August 1935): 88–100.

Corey, Paul. "L'homme du Middle West." Translated by Alex Wittenberg. *Esprit*, n.s., no. 127 (November 1946): 562–567.

Cowley, Malcolm. *Exile's Return: A Literary Odyssey of the 1920s.* New York: Viking Press, 1951. Originally published in 1934.

Cowley, Malcolm. "Naturalism in American Literature." In *American Naturalism*, edited by Harold Bloom, 49–79. Broomall, PA: Chelsea House Publishers, 2004. Originally published in 1950.

Cowley, Malcolm. *Think Back on Us: A Contemporary Chronicle of the 1930's by Malcolm Cowley.* Edited by Henry Dan Piper. Carbondale: Southern Illinois University Press, 1967.

Crapanzano, Vincent. "Life-Histories." *American Anthropologist* 84 (1984): 953–960.

Culver, John C., and John Hyde. *American Dreamer: The Life and Times of Henry A. Wallace.* New York: W. W. Norton, 2000.

Danborn, David B. "Romantic Agrarianism in Twentieth-Century America." *Agricultural History* 65, no. 4 (2000): 1–12.

Denning, Michael. *The Cultural Front: The Laboring of American Culture in the Twentieth Century.* New York: Verso, 1997.

Dennis, Philip A., and Wendell Aycock, eds. *Literature and Anthropology.* Lubbock: Texas Tech University Press, 1989.

Denton, Sally. *The Plots against the President: FDR, A Nation in Crisis, and the Rise of the American Right.* New York: Bloomsbury Press, 2012.

de Pina-Cabral, João. "History of Anthropology and Personal Biography." *Anthropology Today* 24, no. 6 (2008): 26–27.

Doukas, Dimitra. *Worked Over: The Corporate Sabotage of an American Community.* Ithaca, NY: Cornell University Press, 2003.

Dunaway, David King. "Oral Biography." *Biography* 14, no. 3 (1991): 256–266.

Durrenberger, E. Paul. *Anthropology Unbound: A Field Guide to the 21st Century.* New York: Oxford University Press, 2016.

Durrenberger, E. Paul. *The Dynamics of Medieval Iceland: Political Economy and Literature.* Iowa City: University of Iowa Press, 1992.

Durrenberger, E. Paul. "Household Economies and Agrarian Unrest in Iowa—1931." In *Household Economies and Their Transformations*, edited by Morgan Maclaughlan, 198–211. Lanham, MD: University Press of America, 1987.

Durrenberger, E. Paul. "Malinowski Award Lecture: Living Up to Our Words." *Human Organization* 73, no. 4 (2014): 299–304.

Durrenberger, E. Paul. "Notes on the Cultural-Historical Background to the Middlewestern Farm Crisis." *Culture and Agriculture* 28 (Winter 1985/1986): 15–17.

Durrenberger, E. Paul, and Dimitra Doukas. "Class in the USA." *Dialectical Anthropology* 42 (2018): 1–13.

Durrenberger, E. Paul, and Dimitra Doukas. "Gospel of Wealth, Gospel of Work: Hegemony in the U.S. Working Class." *American Anthropologist* 110, no. 2 (2008): 214–224.

Durrenberger, E. Paul, and Suzen Erem. "The Weak Suffer What They Must: A Natural Experiment in Thought and Structure." *American Anthropologist* 101, no. 4 (1999): 783–793.

Dyson, Lowell K. *Farmers' Organizations*. New York: Greenwood Press, 1986.

Dyson, Lowell K. *Red Harvest: The Communist Party and American Farmers*. Lincoln: University of Nebraska Press, 1982.

Eddy, Mary Baker. 1903. *Science and Health with Key to the Scriptures*. Boston: Christian Science Board of Directors.

Eisinger, Chester E. "Character and Self in Fiction on the Left." In *Proletarian Writers of the Thirties*, edited by Harry T. Moore, 158–183. Carbondale: Southern Illinois University Press, 1968.

Eisinger, Chester E. *Fiction of the Forties*. Chicago: University of Chicago Press, 1963.

Elkins, Aaron. *Old Bones*. New York: Mysterious Press, 1987.

Ellsworth, Clayton S. "Ceres and American Men of Letters since 1929." *Agricultural History* 24, no. 4 (1950): 177–181.

Evans-Pritchard, E. E. *The Sanusi of Cyrenaica*. Oxford: Clarendon Press, 1949.

Ewers, John C. "Review of *Ecology and Cultural Continuity as Contributing Factors in the Social Organization of the Plains Indians* by Symmes C. Oliver." *Ethnohistory* 10, no. 1 (1963): 88–90.

Fast, Howard. *Being Red*. Boston: Houghton Mifflin, 1990.

Federal Writers' Project, Works Progress Administration for the State of Iowa. *Iowa: A Guide to the Hawkeye State*. New York: Viking Press, 1938.

Fernia, Elizabeth. *Guests of the Sheik: An Ethnography of an Iraqi Village*. New York: Random House, 1965.

Field, Bruce. *Harvest of Dissent: The National Farmers Union and the Early Cold War*. Lawrence: University Press of Kansas, 1998.

Field, Bruce. "The Price of Dissent: The Iowa Farmers Union and the Early Cold War, 1945–1954." *Annals of Iowa* 55, no. 1 (1996): 1–23.

Fikes, Jay Courtney. *Carlos Castaneda: Academic Opportunism and the Psychedelic Sixties*. Victoria, BC: Millenia Press, 1993.

Flanagen, John T. "The Middle Western Farm Novel." *Minnesota History* 23, no. 2 (1942): 113–125.

Folsom, Franklin. *Days of Anger, Days of Hope: A Memoir of the League of American Writers 1937–1942*. Niwot: University Press of Colorado, 1994.

Fones-Wolf, Elizabeth. *Selling Free Enterprise: The Business Assault on Labor and Liberalism, 1945–1960*. Urbana: University of Illinois Press, 1994.

Frank, Gelya. "Mercy's Children." *Anthropology and Humanism* 6, no. 4 (1981): 8–12.

Frank, Gelya. "Ruth Behar's Biography in the Shadow: A Review of Reviews." *American Anthropologist* 97, no. 2 (1995): 357–374.

Fried, Emanuel. *The Most Dangerous Man*. Kansas City, KS: John Brown Press, 2010.

Friedland, William H. "Who Killed Rural Sociology? A Case Study in the Political Economy of Knowledge Production." *International Journal of Sociology of Agriculture and Food* 17 (2010): 72–88.

Gardiner, John. *The Art of Fiction: Notes on Craft for Young Writers*. New York: Knopf, 1984.

Gardner, Susan. "Introduction." In *Waterlily*, by Ella Cara Deloria, v–xxviii. Lincoln. University of Nebraska Press, 2009.

Gardner, Susan. " 'Though It Broke My Heart to Cut Some Bits I Fancied': Ella Deloria's Original Design for 'Waterlily.' " *American Indian Quarterly* 27, no. 3/4 (2003): 667–696.

Geertz, Clifford. *The Interpretation of Cultures*. New York: Basic Books, 1973.

Geismar, Maxwell. "Naturalism Yesterday and Today." *English Journal* 43, no. 1 (1954): 1–6.

Geismar, Maxwell. *Writers in Crisis: The American Novel, 1925–1940*. Boston: Houghton Mifflin, 1942.

Gerber, Philip. "Becoming an Author . . . 1930s Style." *Books at Iowa* 61 (1994): 7–27.

Glazier, Jack. "Paul Radin and the Fisk University Narrative Collection." *Anthropology News*, https://anthropology-news.org/index.php/2020/02/25/paul-radin-and-the-fisk-university-narrative-collection/. Accessed 06/07/2021.

Glusker, Susannah Joel. *Anita Brenner: A Mind of Her Own*. Austin: University of Texas Press, 1998.

Goldschmidt, Walter. *As You Sow*. Glencoe, IL: The Free Press, 1947.

Goldschmidt, Walter. "Pietro's House." *Anthropology and Humanism* 20, no. 1 (1995): 47–51.

Graham, Ruth. "Mystery Man." Poetry Foundation, September 24, 2013. https://www.poetryfoundation.org/articles/70053/mystery-man. Accessed 2/6/2018.

Greenleaf, Stephen. *Book Case*. New York: William Morrow, 1991.

Greenleaf, Stephen. *Fatal Obsession*. New York: Dial Press, 1983.

Gruber, Jacob. "In Search of Experience." In *Pioneers of American Anthropology: The Uses of Biography*, edited by June Helm, 3–28. Seattle: University of Washington Press, 1966.

Grunfeld, A. Tom, ed. *On Her Own: Journalistic Adventures from San Francisco to the Chinese Revolution 1917–1927*, by Milly Bennett. Armonk, NY: M. E. Sharpe, 1993.

Hallwas, John. "Poetry and Murder: Prairie City's Decker Press." *Macomb Daily Journal*, February 15, 1981. http://www.connectotel.com/patchen/deckjh.html. Accessed 6/6/2021.

Handleman, Don. "Critiques of Anthropology: Literary Turns, Slippery Bends." *Poetics Today* 15, no. 3 (1994): 341–381.

Hans, Birgit. *D'Arcy McNickle: The Hawk Is Hungry and Other Stories*. Tucson: University of Arizona Press, 1992.

Harrison, John M. "Confirmed Typomaniac: Carroll Coleman and the Prairie Press." *Books at Iowa* 62 (1995): 15–74.

Helm, June. "Preface." In *Pioneers of American Anthropology: The Uses of Biography*, edited by June Helm, v–ix. Seattle: University of Washington Press, 1966.

Hemenway, Robert E. *Zora Neale Hurston: A Literary Biography*. Champaign: University of Illinois Press, 1977.

Henry, Jules. *Culture against Man*. New York: Random House, 1963.

Herbst, Josephine. "A Year of Disgrace." In *The Starched Blue Sky of Spain and Other Memoirs*, edited by Diane Johnson, 53–98. New York: Harper Collins, 1991. Originally published in 1960.

Hicks, Granville. *The Great Tradition: An Interpretation of American Literature since the Civil War*. New York: Macmillan, 1933.

Hicks, Granville. *John Reed: The Making of a Revolutionary*. New York: Macmillan, 1936.

Hicks, Granville. *Part of the Truth*. New York: Harcourt, Brace & World, 1965.

Hicks, Granville. *Where We Came Out*. New York: Viking, 1954.

Hochschild, Adam. *Spain in Our Hearts: Americans in the Spanish Civil War, 1936–1939*. Boston, Houghton Mifflin Harcourt, 2016.

Howard, Robert West. *Two Billion Acre Farm: An Informal History of American Agriculture*. Garden City, NY: Doubleday, Doran, 1945.

Huff, Darrell. "Living High on $6500 a Year." *The Saturday Evening Post* (1962): 60–62. Reprinted in *Mother Earth News* (January 1970).

Hughes-Hallett, Lucy. *Gabriele d'Annunzio: Poet, Seducer, and Preacher of War*. New York: Knopf, 2013.

Hugo, Richard. *Death and the Good Life*. New York: St. Martin's Press, 1981.

Hyman, Stanley Edgar. "Constance Rourke and Folk Criticism." *Antioch Review* 7, no. 3 (Autumn 1947): 418–434.

Hynes, James. *Publish and Perish: Three Tales of Tenure and Terror*. New York: Picador, 1997.

Janson, H. W. "The International Aspects of Regionalism." *College Art Journal* 2, no. 4, part 1 (May 1943): 110–115.

Johnson, Diane. "Introduction." In *Josephine Herbst: The Starched Blue Sky of Spain and Other Memoirs*, vii–xxv. New York, Harper Collins, 1991. Originally published in 1960.

Jonsson, Hjorleifur. "Thai Fiction and the Anthropologist." *Cornell Southeast Asia Bulletin* (January 2015): 27–28. https://ecommons.cornell.edu/handle/1813/42442. Accessed 6/7/2021.

Kaplan, Milton A. "American Speech in Radio Poetry." *American Speech* 19, no. 1 (1944): 28–32.

Karr, Rodney D. "Farmer Rebels in Plymouth County, Iowa, 1932–1933." *Annals of Iowa* 47, no. 7 (Winter 1985): 637–645.

Kaysen, Susanna. *Far Afield*. New York: Vintage, 1990.

Kazin, Alfred. *On Native Grounds: An Interpretation of Modern American Prose Literature*. New York: Harcourt, Brace & World, 1942.

Kazin, Alfred. *Starting Out in the Thirties*. Boston: Little, Brown, 1962.

Kessler-Harris, Alice, and Paul Lauter. "Introduction." In *The Unpossessed*, by Tess Slesinger, vii–xvi. New York: Feminist Press, 1984.

King, Charles. *Gods of the Upper Air: How a Circle of Renegade Anthropologists Reinvented Race, Sex, and Gender in the Twentieth Century*. New York: Doubleday, 2019.

King, Florence. *Southern Ladies and Gentlemen*. New York: St. Martin's Press, 1975.

King, Lily. *Euphoria*. New York: Atlantic Monthly Press, 2014.

Kirkendall, Richard S. *Social Scientists and Farm Politics in the Age of Roosevelt*. Columbia: University of Missouri Press, 1966.

Kreymborg, Alfred, Lewis Mumford, and Paul Rosenfeld, eds. *The New Caravan*. Vol. 5. New York: W. W. Norton, 1936.

Kroeber, Alfred. "Introduction." In *American Indian Life: Customs and Traditions of 23 Tribes*, edited by Elsie C. Parsons, 5–16. New York: Dover, 1992. Originally published in 1922.

Kroeber, Theodora. *Ishi in Two Worlds: A Biography of the Last Wild Indian in North America*. Berkeley: University of California Press, 1961.

Kutler, Stanley I. *The American Inquisition: Justice and Injustice in the Cold War*. New York: Hill and Wang, 1982.

La Farge, Oliver. *Behind the Mountains*. Cambridge: Riverside Press, 1951.

La Farge, Oliver. "Foreword." In *Five Families: Mexican Studies in the Culture of Poverty*, by Oscar Lewis, vii–x. New York: Basic, 1962.

La Farge, Oliver. *Laughing Boy: A Navajo Love Story*. New York: Houghton Mifflin, 1929.

La Farge, Oliver. *The Man with the Calabash Pipe: Some Observations*, edited by Winfield Townley Scott. Boston: Houghton Mifflin, 1966.

La Farge, Oliver. *Raw Material*. Boston: Houghton Mifflin, 1945.

Laird, C. G. "Bullet Proof Culture." *Vanity Fair*, November 1931, 58, 59, 80.

Langer, Elinor. "Afterword." In *The Unpossessed*, by Tess Slesinger, 431–449. New York: Feminist Press, 1984.

Langer, Elinor. *Josephine Herbst*. Boston: Little, Brown, 1983.

Langness, L. L., and Gelya Frank. "Fiction and the Ethnographic Novel." *Anthropology and Humanism Quarterly* 3, nos. 1, 2 (1978): 18–22.

Langness, L. L., and Gelya Frank. *Lives: An Anthropological Approach to Biography*. Novato, CA: Chandler & Sharp, 1981.

Langness, L. L., and Gelya Frank. "Mercy's Children." *Anthropology and Humanism* 6, no. 4 (1981): 8–12.

Larson, Olaf F., Robin M. Williams Jr., and Ronald C. Wimberley. "Dismissal of a Sociologist: The AAUP Report on Carl Taylor." *Rural Sociology* 64, no. 4 (1999): 533–553.

Larson, Olaf F., and Julie N. Zimmerman. "The USDA's Bureau of Agricultural Economics and Sociological Studies of Rural Life and Agricultural Issues, 1919–1953." *Agricultural History* 74, no. 2 (2000): 227–240.

Leach, Edmund. *Rethinking Anthropology.* London. Athlone Press, 1961.

Lechlitner, Ruth. ". . . anti-war and anti-fascism . . ." *Carleton Miscellany* 6, no. 1 (1965): 77–82.

Lechlitner, Ruth. *A Changing Season.* Boston: Branden Press, 1973.

Lechlitner, Ruth. "Interview." *New Masses,* July 7, 1936, 20.

Lechlitner, Ruth. "The Last Frontiers." *New Masses,* October 16, 1934, 20.

Lechlitner, Ruth. "Lines for the Year's End." *Poetry* 73, no. 2 (1948): 72–73.

Lechlitner, Ruth. "Ordeal by Tension." *Poetry* 51, no. 6 (1938): 296–299.

Lechlitner, Ruth. "Quiz Program." *Poetry* 58, no. 3 (1941): 132–133.

Lechlitner, Ruth. *The Shadow on the Hour.* Iowa City, IA: Prairie Press, 1956.

Lechlitner, Ruth. *Tomorrow's Phoenix.* New York: Alcestis Press, 1937.

Lechlitner, Ruth. "Wife-Thought." *Sewanee Review* 35, no. 3 (1927): 324.

Lewis, Oscar. "Husbands and Wives in a Mexican Village: A Study of Role Conflict." *American Anthropologist* 51, no. 4 (1949): 602–610.

Lewis, Oscar. *La Vida; A Puerto Rican Family in the Culture of Poverty—San Juan and New York.* New York: Random House, 1966.

Lewis, Oscar. *Life in a Mexican Village: Tepoztlan Restudied.* Urbana: University of Illinois Press, 1951.

Lowie, Robert H. *Robert H. Lowie, Ethnologist: A Personal Record.* Berkeley: University of California Press, 1959.

Lutz, Tom. *Cosmopolitan Vistas: American Regionalism and Literary Value.* Ithaca: Cornell University Press, 2004.

Lynd, Robert S., and Helen M. Lynd. *Middletown: A Study in American Culture.* New York: Harcourt, Brace & World, 1929.

MacLean, Nancy. *Democracy in Chains: The Deep History of the Radical Right's Stealth Plan for America.* New York: Penguin Random House, 2017.

Madden, David. "Introduction." In *Proletarian Writers of the Thirties,* edited by David Madden, xv–xlii. Carbondale: Southern Illinois University Press, 1968.

Malinowski, Bronislaw. *Argonauts of the Western Pacific: An Account of Native Enterprise and Adventure in the Archipelagoes of Melanesian New Guinea.* London: Routledge & Kegan Paul, 1922.

Mangione, Jerre. *The Dream and the Deal: The Federal Writers' Project, 1935–1943.* Boston: Little, Brown, 1972.

Mangione, Jerre. *An Ethnic at Large: A Memoir of America in the Thirties and Forties.* New York, G. P. Putnam's Sons, 1978.

Manning, Richard. "The Trouble with Iowa." *Harper's Magazine,* February 2016, 23–30.

Mayblin, Maya. "The Anarchic Institution." *Anthropology of This Century*, issue 24 (January 2019). http://aotcpress.com/articles/anarchic-institution/. Accessed 1/27/2019.

McCown, Robert A. "Paul Corey's Mantz Trilogy." *Books at Iowa* 17, no. 1 (1972): 15–26.

McNickle, D'Arcy. *Indian Man: A Life of Oliver La Farge*. Bloomington: Indiana University Press, 1971.

Meyer, Michael. "Introduction." In *It Can't Happen Here*, by Sinclair Lewis, v–xv. New York: Signet Classics, 2005.

Meyer, Roy W. *The Middle Western Farm Novel in the Twentieth Century*. Lincoln: University of Nebraska Press, 1965.

Meyer, Roy W. "Naturalism in American Farm Fiction." *Journal of the Central Mississippi Valley American Studies Association* 2, no. 1 (1961): 17–37.

Mickenberg, Julia L. *American Girls in Red Russia: Chasing the Soviet Dream*. Chicago: University of Chicago Press, 2017.

Mighell, A. *A Study of the Organization and Management of Dairy Farms in Northeastern Iowa*. Iowa Agricultural Experiment Station Bulletin 243. Ames: Iowa State College of Agriculture and Mechanic Arts, 1927.

Mills, C. Wright. *Politics, Power, and People: The Collected Essays of C. Wright Mills*, edited by Irving Louis Horowitz. New York: Oxford University Press, 1963.

Mills, C. Wright. *The Power Elite*. New York: Oxford University Press, 1956.

Mills, C. Wright. *The Sociological Imagination*. Oxford: Oxford University Press, 1959.

Miner, Horace. *Culture and Agriculture: An Anthropological Study of a Corn Belt County*. Occasional Contributions from the Museum of Anthropology, no. 14. Ann Arbor: University of Michigan Press, 1949.

Miner, Horace. *The Primitive City of Timbuctoo*. Rev. ed. New York: Doubleday & Company, 1965.

Mintz, Sidney. "The Anthropological Interview and the Life History." In *Oral History: An Interdisciplinary Anthology*, 2nd ed., edited by David K. Dunaway and Willa K., Baum, 298–305. Walnut Creek, CA: AltaMira, 1996.

Mintz, Sidney. "Translator's Note." In *The Vanquished: A Novel*, by Cesar Andreu Iglesias, ix–xv. Chapel Hill: University of North Carolina Press, 2002.

Mitcgang, Herbert. *Dangerous Dossiers: Exposing the Secret War against America's Greatest Authors*. New York: Donald I. Fine, 1988.

Moore, Harry T. "Preface." In *Proletarian Writers of the Thirties*, edited by David Madden, v–vii. Carbondale: Southern Illinois University Press, 1968.

Morrison, D. E., ed. *Farmers' Organizations and Movements: Research Needs and a Bibliography of the United States and Canada*. Agricultural Experiment Station Bulletin 24. East Lansing: Michigan State University, 1970.

Morse, Samuel French. "Review of *Tomorrow's Phoenix*." *Poetry* 51, no. 3 (1937): 157–159.

Munger, H. B. *The Cost of Producing Milk*. Agricultural Experiment Station Bulletin 197. Ames: Iowa State College of Agriculture and Mechanic Arts, 1921.

Munger, H. B. *Iowa Farm Management Surveys in Blackhawk, Grundy, and Tama Counties.* Agricultural Experiment Station Bulletin 198. Ames: Iowa State College of Agriculture and Mechanic Arts, 1921.

Murray, Robert K. *Red Scare: A Study of National Hysteria, 1919–1920.* New York: McGraw-Hill, 1955.

Nanda, Serena, and Joan Gregg. *Assisted Dying: An Ethnographic Murder Mystery on Florida's Gold Coast.* Lanham MD: AltaMira, 2011.

Nanda, Serena, and Joan Gregg. *The Gift of a Bride: A Tale of Anthropology, Matrimony, and Murder.* Lanham, MD: AltaMira, 2009.

Narayan, Kirin. "Ethnography and Fiction: Where Is the Border?" *Anthropology and Humanism* 24, no. 2 (1999): 134–147.

Nekola, Charlotte, and Paula Rabinowitz. *Writing Red: An Anthology of American Women Writers, 1930–1940.* New York: Feminist Press, 1987.

Nelson, Cary. *Repression and Recovery: Modern American Poetry and the Politics of Cultural Memory, 1910–1945.* Madison: University of Wisconsin Press, 1989.

Nelson, Margaret. "My Uncle Paul." *Books at Iowa* 61 (1994): 53–64.

Oliver, Symmes C. "Ecology and Cultural Continuity as Contributing Factors in the Social Organization of the Plains Indians." *University of California Publications in American Archaeology and Ethnology* 48, no. 1 (1962): 1–90.

Olsen, Tillie. *Tell Me a Riddle.* Edited by Deborah Silverton Rosenfelt. New Brunswick, NJ: Rutgers University Press, 1995.

Osanka, Franklin Mark. "Introduction." In *Guerrilla Warfare,* by "Yank" Levy, 7–11. Boulder, CO: Paladin Press, 1964.

Page, Myra. "Preface." In *Daughter of the Hills: A Woman's Part in the Coal Miners' Struggle,* xix–xxiii. New York: Feminist Press, 1977.

Pálsson, Gísli. *The Man Who Stole Himself: The Slave Odyssey of Hans Jonathan.* Chicago: University of Chicago Press, 2016.

Parkin, Frank. *Krippendorf's Tribe.* New York: Antheneum, 1986.

Parrington, Vernon L. *The Beginnings of Critical Realism in America: 1860–1920.* New York: Harcourt, Brace, 1930.

Pells, Richard H. *Radical Visions and American Dreams: Culture and Social Thought in the Depression Years.* New York: Harper & Row, 1973.

Phillips-Fein, Kim. *Invisible Hands: The Making of the Conservative Movement from the New Deal to Reagan.* New York: W. W. Norton, 2009.

Pizer, Donald. "Contemporary American Naturalism." In *American Naturalism,* edited by Harold Bloom, 255–272. Broomall, PA: Chelsea House Publishers, 2004. Originally published in 1993.

Pratt, William C. "Rethinking the Farm Revolt of the 1930s." History Faculty Publications, University of Nebraska, Omaha, Paper 12, 1988. https://digitalcommons .unomaha.edu/histfacpub/12. Accessed 9/22/2017.

Rahv, Philip. "Image and Idea." In *American Naturalism,* edited by Harold Bloom, 39–48. Broomall, PA: Chelsea House, 2004. Originally published in 1949.

Rahv, Philip. "Notes on the Decline of Naturalism." In *American Naturalism*, edited by Harold Bloom, 36–48. Broomall, PA: Chelsea House Publishers, 2004.

Rankin, Nicholas. *Telegram from Guernica: The Extraordinary Life of George Steer, War Correspondent*. London: Faber & Faber, 2003.

Ransom, John Crowe. "The Aesthetics of Regionalism." *American Review* 14 (January 1934): 190–310.

Raup, Philip M. "Corporate Farming in the United States." *Journal of Economic History* 33, no. 1 (March 1973): 274–290.

Redfield, Robert. *Tepoztlan, A Mexican Village: A Study of Folk Life*. Chicago: University of Chicago Press, 1930.

Redfield, Robert, and W. Lloyd Warner. "Cultural Anthropology and Modern Agriculture." In *Farmers in a Changing World: The Yearbook of Agriculture 1940*, edited by Henry A. Wallace, 983–993. Washington, DC: United States Government Printing Office, 1940.

Reid, Panthea. *Tillie Olsen: One Woman, Many Riddles*. New Brunswick, NJ: Rutgers University Press, 2009.

Reigelman, Milton M. *The Midland: A Venture in Literary Regionalism*. Iowa City: University of Iowa Press, 1975.

Rideout, Walter B. *The Radical Novel in the United States 1900–1954: Some Interrelations of Literature and Society*. Cambridge, MA: Harvard University Press, 1956.

Robbin, Natalie. *Alien Ink: The FBI's War on Freedom of Expression*. New York: William Morrow, 1992.

Roberts, Nora Ruth. *Three Radical Women Writers: Class and Gender in Meridel Le Sueur, Tillie Olsen, and Josephine Herbst*. New York: Garland Publishing, 1996.

Rosenfelt, Deborah Silverton. "Introduction." In *Tell Me a Riddle*, by Tillie Olsen, 3–32. New Brunswick, NJ: Rutgers University Press, 1995.

Rourke, Constance. *The Roots of American Culture and Other Essays*. New York: Harcourt, Brace, 1942.

Rush, Norman. *Mating*. New York: Alfred A. Knopf, 1991.

Salzman, Philip Carl. *Understanding Culture: An Introduction to Anthropological Theory*. Prospect Heights, IL: Waveland Press, 2001.

Sawchuk, Kim. "The Cultural Apparatus: C. Wright Mills' Unfinished Work." *American Sociologist* 32, no. 1 (2001): 27–49.

Saxton, Alexander. *Bright Web in the Darkness*. New York: St. Martin's Press, 1958.

Saxton, Alexander. *Grand Crossing*. New York: Harper & Brothers, 1943.

Saxton, Alexander. "Introduction." In *The Great Midland*, xv–xxx. Urbana: University of Illinois Press, 1997.

Saxton, Alexander. *The Great Midland*. New York: Appleton-Century-Crofts, 1948.

Schlesinger, Arthur, Jr. *The Politics of Upheaval: The Age of Roosevelt*. Boston: Houghton Mifflin, 1960.

Schlosser, Eric. *Fast Food Nation: The Dark Side of the All-American Meal*. New York: Harper, 2002.

Schrecker, Ellen. *The Age of McCarthyism: A Brief History with Documents.* New York: St. Martin's Press, 1994.

Schultz, Theodore W. *Agriculture in an Unstable Economy.* New York: McGraw-Hill, 1945.

Scott, James C. *Seeing Like a State: How Certain Schemes to Improve the Human Condition Have Failed.* New Haven, CT: Yale University Press, 1998.

Seabrook, William. "Pioneer Spirit, '39." *Readers Digest* 35, no. 209 (September 1939): 42–46.

Shockley, Martin Staples. "The Reception of *The Grapes of Wrath* in Oklahoma." *American Literature* 15, no. 4 (January 1944): 351–361.

Shover, John L. *Cornbelt Rebellion: The Farmers' Holiday Association.* Urbana: University of Illinois Press, 1965.

Simeone, William E. "Fascists and Folklorists in Italy." *Journal of American Folklore* 91, no. 359 (1978): 543–557.

Solotaroff, Robert. *Robert Stone.* New York: Twayne, 1994.

Stein, Julia. "Industrial Music: Contemporary American Working-Class Poetry and Modernism." *Women's Studies Quarterly* 23, no. 1/2 (1995): 229–247.

Stone, Robert. *A Flag for Sunrise.* New York: Alfred A. Knopf, 1981.

Strong, Anna Louise. *My Native Land.* New York: Viking, 1940.

Strong, Tracy B., and Helene Keyssar. *Right in Her Soul: The Life of Anna Louise Strong.* New York: Random House, 1983.

Suckow, Ruth. "The Folk Idea in American Life." *Scribner's Magazine* 88 (September 1930): 245–255.

Suckow, Ruth. *A Memoir.* New York: Rinehart & Company, 1952.

Susman, Warren I. *Culture as History: The Transformation of American Society in the Twentieth Century.* New York: Pantheon, 1973.

Susman, Warren I., ed. *Culture and Commitment: 1929–1945.* New York: George Braziller, 1973.

Taylor, C. C., and E. G. Hurd. *Farm Organization and Farm Profits in Tama County, Iowa.* Agricultural Experiment Station Bulletin 88. Ames: Iowa State College of Agriculture and Mechanic Arts, 1925.

Taylor, H. C., and A. D. Taylor. *The Story of Agricultural Economics in the United States, 1840–1932.* Ames: Iowa State College Press, 1952.

Temkin, Moshik. "Sacco & Vanzetti Today." *The Nation,* August 27, 2007. https://www.thenation.com/article/sacco-amp-vanzetti-today/. Accessed 1/17/2020.

Thomas, H. L., and J. A. Hopkins Jr. *Costs and Utilization of Corn in Seven Iowa Counties.* Agricultural Experiment Station Bulletin 289. Ames: Iowa State College of Agriculture and Mechanic Arts, 1932.

Thompson, Lawrance, and R. H. Winnick. *Robert Frost: A Biography.* New York: Holt, Rinehart and Winston, 1981.

Tomar, Ruchika. *A Prayer for Travelers.* New York: Random House, 2019.

Trachktenberg, Alan. *The Incorporation of America: Culture and Society in the Gilded Age*. New York: Hill and Wang, 1982.

Tuan, Yi-Fu. *The Good Life*. Madison: University of Wisconsin Press, 1986.

Turton, Michael. "80 Years Later, Author Paul Corey's Influence Still Felt in Philipstown." *Highlands Current*, May 30, 2011. http://highlandscurrent.com/2011/05/30/80-years-later-author-paul-coreys-influence-still-felt-in-philipstown/. Accessed 11/6/2017.

Van Doren, Carl. *Contemporary American Novelists 1900–1920*. New York: Macmillan, 1922.

Vogeler, Ingolf. *The Myth of the Family Farm: Agribusiness Dominance of United States Agriculture*. Boulder, CO: Westview, 1981.

Wald, Alan M. *American Night: The Literary Left in the Era of the Cold War*. Chapel Hill: University of North Carolina Press, 2012.

Wald, Alan M. *Exiles from a Future Time: The Forging of the Mid-Twentieth-Century Literary Left*. Chapel Hill: University of North Carolina Press, 2002.

Wald, Alan M. *Trinity of Passion: The Literary Left and the Antifascist Crusade*. Chapel Hill: University of North Carolina Press, 2007.

Waldman, Amy. *A Door in the Earth*. New York: Little, Brown, 2019.

Wallace, Henry A. "Foreword." In *Farmers in a Changing World: The Yearbook of Agriculture 1940*, edited by Henry A. Wallace, v. Washington, DC: United States Government Printing Office, 1940.

Wallace, Henry A. *Yearbook of Agriculture*. Washington, DC: US Government Printing Office, 1934.

Wallace, Henry A. "The Year in Agriculture: The Secretary's Report to the President." In *Yearbook of Agriculture*, edited by Milton Eisenhower, 1–99. Washington DC: US Government Printing Office, 1934.

Ware, Susan. *Holding Their Own: American Women in the 1930s*. Boston: Twayne Publishers, 1982.

Weiner, John. "Review of *The Cultural Front*." *Reviews in American History* 25, no. 4 (1997): 625–630.

Wertsch, Douglas Michael. "Resisting the Wave: Rural Iowa's War against Crime, 1920–1941." PhD diss., Department of History, Iowa State University, Ames, 1992.

White, Jenny. *The Abyssinian Proof*. New York: W. W. Norton, 2008.

White, Jenny. *Islamist Mobilization in Turkey: A Study in Vernacular Politics*. Seattle: University of Washington Press, 2002.

White, Jenny. *Money Makes Us Relatives: Women's Labor in Urban Turkey*. Austin: University of Texas Press, 1994.

White, Jenny. *Muslim Nationalism and the New Turks*. Princeton, NJ: Princeton University Press, 2014.

White, Jenny. "Portrait of the Scholar as a Young Novelist." *Inside Higher Ed*, February 9, 2006. https://www.insidehighered.com/views/2006/02/09/portrait-scholar-young-novelist. Accessed 12/22/2018.

White, Jenny. *The Sultan's Seal*. New York: W. W. Norton, 2006.

White, Jenny. *The Winter Thief*. New York: W. W. Norton, 2010.

White, Ronald A. *Milo Reno: Farmers Union Pioneer*. New York: Arno Press, 1975.

Whitehead, Ella. 1990. *John Lehmann's "New Writing": An Author Index 1936–1950*. Lewiston, NY: Edwin Mellen Press.

Whiting, Cécile. *Antifascism in American Art*. New Haven, CT: Yale University Press, 1989.

Wilson, Carter. "Introduction." In *Crazy February: Death and Life in the Mayan Highlands of Mexico*, 1–7. Berkeley: University of California Press, 1974.

Wilson, Carter. "Introduction." In *A Green Tree and a Dry Tree: A Novel of Chiapas*, xiii–xxii. Albuquerque: University of New Mexico Press, 1995.

Winks, Robin W. *Cloak and Gown: Scholars in the Secret War, 1939–1961*. New York: William Morrow, 1987.

Winnick, R. H. "Preface." In *Robert Frost: The Later Years, 1938–1963*, by Lawrance Thompson and R. H. Winnick, xi–xiii. New York: Holt, Rinehart and Winston, 1966.

Wintringham, T. H. "Introduction." In *Guerrilla Warfare*, by Bert "Yank" Levy, 5–10. Penguin Special S102. Harmondsworth, UK: Penguin Books, 1941.

Wixson. Douglas. "Looking for Paul Corey: Memory's Loss and the Fall of Self-Reliance." *North American Review* 288, no. 3/4 (2003): 62–70.

Wixson. Douglas. *Worker-Writer in America: Jack Conroy and the Tradition of Midwestern Literary Radicalism, 1898–1990*. Urbana: University of Illinois Press, 1994.

Wolfskill, George. *The Revolt of the Conservatives: A History of the American Liberty League, 1934–1940*. Boston: Houghton Mifflin, 1962.

Wolfskill, George, and John A. Hudson. *All but the People: Franklin D. Roosevelt and His Critics, 1933–1939*. New York: Macmillan, 1969.

Wood, Donald C. *And If Strangers Come to Supper*. Amazon/CreateSpace, 2015. https://www.amazon.com/dp/B00X1F3U1M/ref=dp-kindle-redirect?_encoding=UTF8&btkr=1. Accessed 6/7/2021.

Wood, Donald C. *Ogata-Mura: Sowing Dissent and Reclaiming Identity in a Japanese Farming Village*. New York: Berghahn Books, 2012.

Wood, Grant. *Revolt against the City*. Whirling World Series, no. 1. Iowa City, IA: Clio Press, 1935.

Woodard, Colin. *American Nations: A History of the Eleven Rival Regional Cultures of North America*. New York: Penguin, 2011.

INDEX